Lecture Notes in Computer Science 16566

Founding Editors

Gerhard Goos
Juris Hartmanis

Editorial Board Members

Elisa Bertino, *Purdue University, West Lafayette, IN, USA*
Wen Gao, *Peking University, Beijing, China*
Bernhard Steffen, *TU Dortmund University, Dortmund, Germany*
Moti Yung, *Columbia University, New York, NY, USA*

The series Lecture Notes in Computer Science (LNCS), including its subseries Lecture Notes in Artificial Intelligence (LNAI) and Lecture Notes in Bioinformatics (LNBI), has established itself as a medium for the publication of new developments in computer science and information technology research, teaching, and education.

LNCS enjoys close cooperation with the computer science R & D community, the series counts many renowned academics among its volume editors and paper authors, and collaborates with prestigious societies. Its mission is to serve this international community by providing an invaluable service, mainly focused on the publication of conference and workshop proceedings and postproceedings. LNCS commenced publication in 1973.

Gustavo Carvalho · Tsutomu Kobayashi
Editors

Formal Methods Teaching

7th Formal Methods Teaching Workshop, FMTea 2026
Tokyo, Japan, May 19, 2026
Proceedings

 Springer

Editors
Gustavo Carvalho ⓘ
Universidade Federal de Pernambuco
Recife, Brazil

Tsutomu Kobayashi ⓘ
Japan Aerospace Exploration Agency
Ibaraki, Japan

ISSN 0302-9743 ISSN 1611-3349 (electronic)
Lecture Notes in Computer Science
ISBN 978-3-032-26742-9 ISBN 978-3-032-26743-6 (eBook)
https://doi.org/10.1007/978-3-032-26743-6

This Springer imprint is published by the registered company Springer Nature Switzerland AG
The registered company address is: Gewerbestrasse 11, 6330 Cham, Switzerland

If disposing of this product, please recycle the paper.

Preface

Formal Methods are regarded as essential for establishing software development as a mature engineering discipline. Today, formal methods thrive in niche areas of industry, especially where safety and security are of critical importance, as illustrated, for example, by the FM 2026 Industry Day. These applications are typically driven by highly skilled and dedicated practitioners. However, although everyday software development could equally benefit from formal methods, post-secondary education in software development has not yet fully achieved this goal.

Many questions remain open: Which formalisms should be taught in dedicated courses, and which should be integrated into existing ones? Which formal tools should be introduced early in undergraduate curricula, and which are more appropriate at the graduate level? Is it more effective pedagogically to integrate formal methods and their associated tools with existing languages and environments, or to teach them separately? How can students be motivated to learn formal methods if they are not widely demanded by employers? What principles and tools can support the teaching and assessment of large classes using formal methods? The FMTea Workshop Series (https://fmtea.github.io), organized by the Teaching Committee of Formal Methods Europe (https://fme-teaching. github.io), aims to address these and related questions.

The FMTea 2026 Workshop took place on May 19, 2026, in Tokyo, Japan, as part of the 27th International Symposium on Formal Methods (FM 2026). The program featured two invited talks. The first was given by Alcino Cunha (INESC TEC and University of Minho, Portugal) on "Teaching Logic with Specification Challenges." The second was delivered by Jim Woodcock (Southwest University, China; Aarhus University, Denmark; University of York, UK) on "Teaching Unifying Theories of Programming." In addition, the program included six regular papers. Out of 12 submissions, each reviewed by three Program Committee members in a single-blind process, six were selected for presentation and inclusion in these proceedings.

We would like to express our sincere gratitude to the FME Teaching Committee and, in particular, its Chair, Luigia Petre (Åbo Akademi University, Finland), for her guidance; to the Program Committee members for their dedication; to the EasyChair organization for providing their technical infrastructure; to Springer for publishing the proceedings in the Formal Methods Teaching series (https://link.springer.com/confer ence/tfm); and to the General Chair of FM 2026, Fuyuki Ishikawa (National Institute of Informatics, Japan), as well as the Workshop Chairs, Elena Troubitsyna (KTH, Sweden) and Tomoyuki Yokogawa (Okayama Prefectural University, Japan), for their valuable organizational support.

May 2026

Gustavo Carvalho
Tsutomu Kobayashi

Organization

Program Committee Chairs

Gustavo Carvalho   Universidade Federal de Pernambuco, Brazil
Tsutomu Kobayashi   Japan Aerospace Exploration Agency, Japan

Program Committee

Erika Abraham   RWTH Aachen University, Germany
Sandrine Blazy   University of Rennes - IRISA, France
Gustavo Carvalho   Universidade Federal de Pernambuco, Brazil
Brijesh Dongol   University of Surrey, UK
Catherine Dubois   ENSIIE-Samovar, France
Tsutomu Kobayashi   Japan Aerospace Exploration Agency, Japan
Thierry Lecomte   CLEARSY, France
Michael Leuschel   University of Düsseldorf, Germany
Tim Nelson   Brown University, USA
Marcel Oliveira   Universidade Federal do Rio Grande do Norte, Brazil
David Pearce   Victoria University of Wellington, New Zealand
Luigia Petre   Åbo Akademi University, Finland
Leila Ribeiro   Universidade Federal do Rio Grande do Sul, Brazil
Pierluigi San Pietro   Politecnico di Milano, Italy
Emil Sekerinski   McMaster University, Canada
Graeme Smith   The University of Queensland, Australia

Contents

Invited Talks

Teaching Logic with Specification Challenges

Alcino Cunha$^{(\boxtimes)}$ and Nuno Macedo

INESC TEC and University of Minho, Braga, Portugal
`{alcino,nmacedo}@di.uminho.pt`

Abstract. Alloy is a lightweight formal method that is well suited to teaching logic because it combines expressive logics with automatic analysis and visual feedback. In this paper, we report our experience using specification challenges on the Alloy4Fun platform in our formal methods courses. We briefly describe the types of challenges we have used over the years and discuss how different hints can help students make progress in solving them. Our main conclusion is that specification challenges are highly engaging and useful for students, but they should be balanced with broader modeling and validation activities to support long-term learning outcomes. They are also useful for research, because Alloy4Fun's data-collection infrastructure enables the release of open datasets that can be mined for insights into Alloy usage and for the evaluation of new tools and techniques.

Keywords: Formal methods · Education · Alloy · Specification challenges · Hints · Open data

1 Teaching Logic with Alloy

We have been teaching state-based formal methods at University of Minho since the mid-1990s. In 2001, we began teaching Alloy [12,13], a formal specification language proposed by Daniel Jackson at MIT and originally designed to reason about software structures. In Alloy, structures are modeled using the unifying concept of a mathematical relation, and constraints about those structures are specified using *Relational Logic* (RL), an extension of *First-Order Logic* (FOL) with relational algebra and closure operators. Alloy is well suited for teaching logic because its Analyzer can automatically verify assertions and check the consistency of assumptions. This automated analysis is possible because a *scope* is imposed to limit the size of the domain. Moreover, the Analyzer can enumerate and graphically depict instances (showing that assumptions are consistent) or counterexamples (showing that assertions are invalid), which is appealing to users and can help them debug and improve specifications.

In 2016, we proposed an extension of Alloy to simplify the specification and analysis of behavioral requirements [18]. In this extension, relations can be declared as mutable, and *Linear Temporal Logic* (LTL) operators can be

G. Carvalho and T. Kobayashi (Eds.): FMTea 2026, LNCS 16566, pp. 3–11, 2026.
https://doi.org/10.1007/978-3-032-26743-6_1

used in specifications. It also supports the *prime* operator from the *Temporal Logic of Actions* (TLA) [16]. This allows a transition system to be specified using a standard temporal-logic formula, without the need for a domain-specific language. Analysis is performed natively via SAT-based bounded model checking, or via complete model checking by translation to off-the-shelf model checkers [17]. Iteration and visual depiction of instance and counterexample traces are still supported. In 2021, this extension officially became Alloy version 6. As a result, Alloy is now an excellent formal method for teaching FOL and LTL, the most common logics for specifying structural and behavioral requirements, respectively.

2   A Brief History of Alloy4Fun

Inspired by Microsoft's Rise4Fun initiative [2], and in particular by Pex4Fun [27] (both now inactive), we started developing Alloy4Fun—a web application for Alloy—around 2015. Rise4Fun was a web service that allowed formal-methods and programming-language researchers to easily deploy their tools on the web. Pex4Fun allowed students to engage in coding duels in which they attempted to guess a secret method implementation (returning unit tests as counterexamples if the implementation was not equivalent). We wanted Alloy4Fun to support some of the core features of Rise4Fun and Pex4Fun, namely: easy sharing of Alloy models and specifications via permalinks; collection of models and specifications to build a valuable corpus of data to support research; and support for specification duels (or challenges) in which students tried to guess a secret specification. However, we also wanted additional features that prevented us from deploying Alloy4Fun in the Rise4Fun web service, namely: visual depiction of instances and counterexamples, as in the Alloy Analyzer; easy sharing of these instances and counterexamples via permalinks (preserving layout and theme); a more refined data-collection mechanism (more on this later); and fully anonymous interaction (to keep interaction simple and avoid privacy and security issues).

The first version of Alloy4Fun was ready in 2016 [24], and our first experiments using specification challenges to teach logic in graduate-level courses began in 2018. With feedback from these early experiments, we refined the application and started using it more systematically in 2019. In 2020, we presented Alloy4Fun at ABZ [21] and released the first Zenodo open dataset [19]. An extended version of the ABZ paper was published in 2021 [22]. Later, we gave tutorials on Alloy4Fun at ABZ in 2024 and in the FMTea tutorial series in 2024. In 2025, we released the latest version of the Zenodo open dataset [20]. Alloy4Fun is currently deployed at http://alloy4fun.inesctec.pt, but it is open source and can be deployed locally. More information about Alloy4Fun is available at https:// haslab.github.io/Alloy4Fun/.

3   Specification Challenges

Alloy4Fun allows a paragraph of a model to be marked as secret using the special comment `//SECRET`. If a model containing secrets is shared, two permalinks are generated: one that provides access only to the public content and another that provides access to the full model. Instructors can use this special comment to create simple specification challenges, as shown in Fig. 1. This example is a (very abstract) model of a social network. It declares two signatures: User, the set of users, and Influencer, the subset of users who are influencers. It also declares a binary relation follows that associates each user to the users they follow. We then have the public (empty) predicate Spec, in which the student is expected to formalize a requirement (given in natural language in a comment), followed by the secret predicate Oracle, which contains a correct formalization of that requirement. Finally, we have a secret command that checks whether the specification provided by the student is equivalent to the one in the oracle. A scope of 4 is used, meaning that the command is checked for all instances with up to 4 users. This challenge can be accessed at http://alloy4fun.inesctec.pt/ YZCN5i2J78veQJ6um. We now briefly describe the different types of challenges we used over the years.

```
sig User { follows : set User }
sig Influencer extends User {}
pred Spec { // Influencers are followed by everyone else

}
//SECRET
pred Oracle { all u : Influencer | follows.u = User - u }
//SECRET
check Spec { Spec iff Oracle } for 4
```

Fig. 1. A simple specification challenge.

Structural Requirements. Most of our challenges aim to practice RL (and FOL) and ask for simple structural requirements over small domain models, similar to the one in Fig. 1. Usually, we list multiple challenges in a single exercise, targeting different requirements of the same domain model. These requirements are cumulative: when attempting a challenge, the student can assume that all previous requirements are satisfied (even if they did not specify them correctly, or at all). To achieve this, the command that checks equivalence to the oracle is an implication whose left-hand side is the conjunction of all previous oracles. One such exercise can be found at http://alloy4fun.inesctec.pt/u6EPDxzJrQSEKRwSf; it is an evolution of our example challenge, with a more detailed domain model, and it is the most attempted exercise by students using Alloy4Fun.

Behavioral Requirements. We also proposed challenges aimed at practicing LTL. Initially, we asked for arbitrary requirements that referred to the various mutable structures of a model. However, these challenges still required strong knowledge of RL to solve, rather than focusing mainly on LTL. Inspired by the *operational principles* of Daniel Jackson's *software concepts* [14], we later evolved to challenges in which the mutable state was opaque and we only provided students with the interface of the actions available to mutate it. The requirements then referred only to the possible orders in which actions could occur. An example of one such challenge can be found at http://alloy4fun.inesctec.pt/M65cdRJE4Jci2nnKY. Specifying these challenges in Alloy4Fun is a bit more cumbersome, as can be seen at http://alloy4fun.inesctec.pt/GxKTndgdDTxewzX8X.

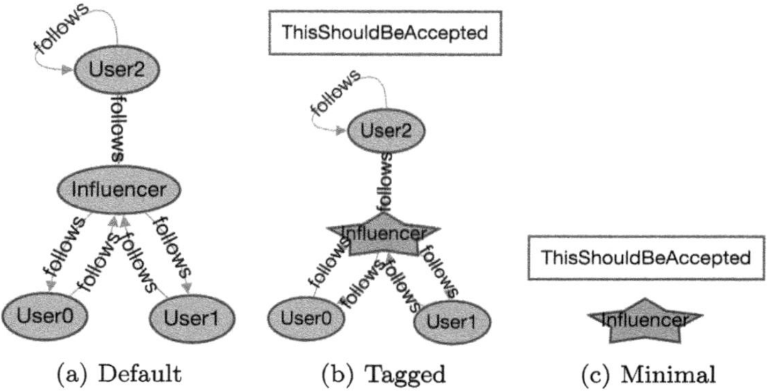

(a) Default   (b) Tagged   (c) Minimal

Fig. 2. Counterexamples.

Actions. To practice specifying state-transition systems, we also proposed challenges in which students were asked to specify the guards and effects of actions. Initially, we used challenges in which students were asked to specify the full transition system at once, but these were very difficult to debug and solve. We now ask for the specification of a single action at a time. However, when checking equivalence to the oracle, this requires assuming that the pre-state satisfies an invariant characterizing all reachable states of the transition system, which might not always be trivial to obtain. These challenges require only knowledge of RL and the prime operator. One example can be found at http://alloy4fun. inesctec.pt/SEYtemwhLRTAzLZEP.

4   Helping Students with Hints

The counterexamples returned by Alloy already serve as a hint to help students debug and fix their specifications. For example, if a student specifies the requirement from Fig. 1 incorrectly as **all** u : Influencer | follows.u = User, a possible counterexample is the one shown in Fig. 2a. However, anecdotal evidence suggests that students struggle to understand even such simple counterexamples, so we began investigating how to provide better hints to students.

Better Visualizations. One way to help students understand counterexamples is to improve their visualizations. As in the Alloy Analyzer, Alloy4Fun allows us to customize the visualization theme (e.g., by assigning different shapes or colors to different entities). When a model (or challenge) is shared, this theme is preserved, so students can immediately see counterexamples rendered using the theme designed by the instructor. Alloy4Fun's visualization component could still be improved, particularly for temporal scenarios, where it is far from ideal. Several works have proposed improvements to Alloy's visualization mechanism (e.g., [6,10,25]) that could, in principle, be reused in Alloy4Fun.

Explanatory Tags. Students sometimes have difficulty determining whether a counterexample satisfies the requirement. This information is important because it helps them fix the specification: if it does, they must relax their specification; if it does not, they must strengthen it. We can use Alloy itself to add explanatory tags to counterexamples, and many of our challenges now include tags indicating whether a counterexample "should be rejected" or "should be accepted." For an example of how this can be implemented, see http://alloy4fun. inesctec.pt/AP49TXH9X3i53zGt8. Together with an improved theme, such tags have proven popular with our students. If you visit http://alloy4fun.inesctec.pt/ Dpm8focEwXZ3Jk3cR and enter the incorrect specification above, the previous counterexample will be depicted as in Fig. 2b. We have used similar tags for other purposes, such as showing progress (e.g., increasing grades) in challenges that require larger specifications.

Better Counterexamples. Another way to provide better hints is to show counterexamples that reduce cognitive effort—for example, minimal ones, or counterexamples that are as close as possible to those seen before fixing a specification. The minimal counterexample for our incorrect specification is depicted in Fig. 2c. By default, Alloy does not allow control over the order in which counterexamples are enumerated. However, several extensions of Alloy have been proposed to address this (e.g. [9,23,29]). For now, we just use the scope to remove irrelevant entities and unclutter the counterexamples somewhat.

Precise Error Locations. Rather than easing the understanding of counterexamples, we could point the student directly to the part of the formula that is wrong. Many automatic specification repair techniques have been developed for Alloy (e.g., [1,3–5,11,28,30]), which could be leveraged to provide such precise error locations. We have developed a prototype of Alloy4Fun that supports this feature, based on a specification repair technique of our own [5]. For example, given the incorrect specification **all** `u : Influencer | follows.u = User`, it would suggest that the user try changing the expression `User`.

Natural Language Hints. The increasing efficacy of *Large Language Models* (LLMs) opens exciting new possibilities, such as providing hints in natural language. These could, for example, explain to the student why a counterexample should be accepted or rejected, or indicate which requirement their incorrect specification is actually formalizing (rather than the intended one).

Do Any of These Have an Impact? So far, the only hints we implemented in the Alloy4Fun challenges used in our courses are improved visualization themes and explanatory tags. Before deploying other kinds of hints, we decided to conduct a large empirical user study to assess whether different hints actually affect immediate performance and learning retention [8]. We compared three kinds of hints—minimal counterexamples with explanatory tags, precise error locations, and a natural language description of the incorrect requirement that is actually being formalized—and concluded that none of them has a significant impact on learning retention, and that only precise error locations have an impact on immediate performance. In other words, some hints may help during study, but they do not impact students' later ability to specify requirements without hints.

5   Collecting and Mining Data

As mentioned above, Alloy4Fun collects data in the background. Every time an analysis command is run, it stores the corresponding model and result, as well as a pointer to the parent model from which it was derived. This data can be mined for relevant statistics and insights. Anyone in possession of a private permalink (typically an instructor who created a specification challenge) can download the data for all models in the derivation tree of the corresponding public model (which contains all attempts to solve a challenge). We also precompute statistics that can be used to quickly identify the most problematic challenges and the most frequent errors. Instructors can inspect these statistics to identify possible learning breakdowns and revise concepts accordingly.

We have also made public the data collected from many of the challenges we have proposed in our formal-methods courses over the years. This data is available as an open dataset on Zenodo [20] and has already been used to evaluate many techniques and tools for Alloy, including most of the automatic specification repair techniques mentioned above, as well as a recent LLM-based technique for generating scenarios to validate formal specifications [7]. It has also been mined to gain insights into how students learn and specify in Alloy, for instance by identifying the most common mistakes. Our analysis of the first Alloy4Fun dataset [22] showed that some features of Alloy's logic are indeed harder for students to master (e.g., nested quantifiers or binary temporal operators). A more extensive, more recent study of the Alloy4Fun dataset can be found in [15].

6   Conclusion

In hindsight, perhaps the biggest benefit of the Alloy4Fun specification challenges was for us, as researchers, since the collected data enabled the evaluation of interesting new techniques and tools. Owing to the gamification aspect, students enjoy the specification challenges and use them frequently while studying, but we are not entirely sure whether they are beneficial in the long term. In particular, in some academic years in which we relied too heavily on these challenges, we noted a decline in students' ability to develop and validate models from scratch. As a result, we have recently scaled back the use of specification challenges in class and introduced more modeling and specification exercises to be developed in the classic Alloy Analyzer. We also realized that we need to teach more validation techniques, in particular by adopting a test-driven modeling approach in which students are encouraged to first come up with a good test suite that clarifies possible misconceptions about a requirement [26]. For the requirements we previously used in challenges, we can actually mine the data to identify the most frequent misconceptions and perhaps propose complementary "validation challenges" in Alloy4Fun.

Acknowledgments. This work is funded by national funds through FCT - Fundação para a Ciência e a Tecnologia, I.P., under the support UID/50014/2025 (https://doi.org/10.54499/UID/50014/2025). We used AI to proofread and improve the grammar of this paper using the Prism workspace provided by OpenAI, which integrates directly with GPT-5.2.

References

1. Alhanahnah, M., Rashedul Hasan, M., Xu, L., Bagheri, H.: An empirical evaluation of pre-trained large language models for repairing declarative formal specifications. Empir. Softw. Eng. **30**(5), 149 (2025)
2. Ball, T., de Halleux, P., Swamy, N., Leijen, D.: Increasing human-tool interaction via the web. In: 11th ACM SIGPLAN-SIGSOFT Workshop on Program Analysis for Software Tools and Engineering, pp. 49–52 (2013)
3. Barros, A., Neto, H., Cunha, A., Macedo, N., Paiva, A.C.: Alloy repair hint generation based on historical data. In: 26th International Symposium on Formal Methods, pp. 104–121. Springer (2024). https://doi.org/10.1007/978-3-031-71177-0_8
4. Brida, S.G., et al.: BeAFix: an automated repair tool for faulty Alloy models. In: 36th IEEE/ACM International Conference on Automated Software Engineering, pp. 1213–1217. IEEE (2021)
5. Cerqueira, J., Cunha, A., Macedo, N.: Timely specification repair for Alloy 6. In: 20th International Conference on Software Engineering and Formal Methods, pp. 288–303. Springer (2022). https://doi.org/10.1007/978-3-031-17108-6_18
6. Couto, R., Campos, J.C., Macedo, N., Cunha, A.: Improving the visualization of Alloy instances. In: 4th Workshop on Formal Integrated Development Environment, pp. 37–52 (2018)
7. Cunha, A., Macedo, N.: Validating formal specifications with LLM-generated test cases. In: 27th International Symposium on Formal Methods. Springer (2026), to appear

8. Cunha, A., Macedo, N., Campos, J.C., Margolis, I., Sousa, E.: Assessing the impact of hints in learning formal specification. In: 46th International Conference on Software Engineering: Software Engineering Education and Training, pp. 151–161 (2024)

9. Cunha, A., Macedo, N., Guimarães, T.: Target oriented relational model finding. In: 17th International Conference on Fundamental Approaches to Software Engineering, pp. 17–31. Springer (2014). https://doi.org/10.1007/978-3-642-54804-8_2

10. Dyer, T., Baugh, J.: Sterling: A web-based visualizer for relational modeling languages. In: 8th International Conference on Rigorous State-Based Methods, pp. 99–104. Springer (2021). https://doi.org/10.1007/978-3-030-77543-8_7

11. Gutiérrez Brida, S., et al.: ICEBAR: feedback-driven iterative repair of Alloy specifications. In: 37th IEEE/ACM International Conference on Automated Software Engineering, pp. 1–13 (2022)

12. Jackson, D.: Alloy: a lightweight object modelling notation. ACM Trans. Softw. Eng. Methodol. **11**(2), 256–290 (2002)

13. Jackson, D.: Software Abstractions: Logic, Language, and Analysis. MIT Press (2012)

14. Jackson, D.: The Essence of Software: Why Concepts Matter for Great Design. Princeton University Press (2021)

15. Jovanovic, A., Sullivan, A.: Right or wrong–understanding how users write software models in Alloy. In: 22nd International Conference on Software Engineering and Formal Methods, pp. 309–327. Springer (2024). https://doi.org/10.1007/978-3-031-77382-2_18

16. Lamport, L.: The temporal logic of actions. ACM Trans. Program. Lang. Syst. **16**(3), 872–923 (1994)

17. Macedo, N., Brunel, J., Chemouil, D., Cunha, A.: Pardinus: a temporal relational model finder. J. Autom. Reason. **66**(4), 861–904 (2022)

18. Macedo, N., Brunel, J., Chemouil, D., Cunha, A., Kuperberg, D.: Lightweight specification and analysis of dynamic systems with rich configurations. In: 24th International Symposium on Foundations of Software Engineering, pp. 373–383 (2016)

19. Macedo, N., Cunha, A., Paiva, A.C.R.: Alloy4Fun dataset for 2019/20 (2020). https://doi.org/10.5281/zenodo.4665672

20. Macedo, N., Cunha, A., Paiva, A.C.R.: Alloy4Fun dataset for 2024/25 (2025). https://doi.org/10.5281/zenodo.17390557

21. Macedo, N., et al.: Experiences on teaching Alloy with an automated assessment platform. In: 7th International Conference on Rigorous State-Based Methods, pp. 61–77. Springer (2020). https://doi.org/10.1007/978-3-030-48077-6_5

22. Macedo, N., et al.: Experiences on teaching Alloy with an automated assessment platform. Sci. Comput. Program. **211**, 102690 (2021)

23. Nelson, T., Saghafi, S., Dougherty, D.J., Fisler, K., Krishnamurthi, S.: Aluminum: principled scenario exploration through minimality. In: 35th International Conference on Software Engineering, pp. 232–241 (2013)

24. Pereira, J.: A web-based social environment for Alloy. Master's thesis, Universidade do Minho (2016)

25. Prasad, S., Greenman, B., Nelson, T., Krishnamurthi, S.: Lightweight diagramming for lightweight formal methods: a grounded language design. In: 39th European Conference on Object-Oriented Programming, pp. 26–1 (2025)

26. Prasad, S., Greenman, B., Nelson, T., Wrenn, J., Krishnamurthi, S.: Making hay from wheats: a class sourcing method to identify misconceptions. In: 22nd Koli Calling International Conference on Computing Education Research, pp. 1–7 (2022)

27. Tillmann, N., De Halleux, J., Xie, T., Bishop, J.: Pex4Fun: teaching and learning computer science via social gaming. In: 25th Conference on Software Engineering Education and Training, pp. 90–91. IEEE (2012)
28. Wang, K., Sullivan, A., Khurshid, S.: Arepair: a repair framework for Alloy. In: 41st International Conference on Software Engineering: Companion Proceedings, pp. 103–106. IEEE (2019)
29. Zhang, C., et al.: AlloyMax: bringing maximum satisfaction to relational specifications. In: 29th ACM Joint Meeting on European Software Engineering Conference and Symposium on the Foundations of Software Engineering, pp. 155–167 (2021)
30. Zheng, G., et al.: ATR: template-based repair for alloy specifications. In: 31st ACM SIGSOFT International Symposium on Software Testing and Analysis, pp. 666–677 (2022)

Teaching Unifying Theories of Programming

Jim Woodcock[1,2,3]($\boxtimes$)

[1] Southwest University, Chongqing, China
[2] Aarhus University, Aarhus, Denmark
[3] University of York, York, UK
jim.woodcock@york.ac.uk
https://www-users.york.ac.uk/~jw524/

Abstract. Teaching Unifying Theories of Programming (UTP) exposes a recurring problem in formal methods education: realistic semantic models are rich, but students learn only when they can see a clear next step. Our central claim is that UTP is most effectively taught not primarily as a semantic meta-theory, but as a disciplined method for specifying, calculating, and refining programs. In this *FMTea* keynote paper, we argue that we should organise UTP teaching around three linked unifications: across theory families, across abstraction levels, and across semantic presentations. With this organisation, we give students a stable route through the subject. They learn one core relational and refinement calculus, then reuse it across designs, reactive contracts, linked models, and beyond. They connect requirements to implementations through explicit refinement steps, and they view denotational, operational, and algebraic accounts not as competing definitions but as mutually justifying views of the same theory. The paper presents a course architecture centred on a minimal teachable core, and a taxonomy of exercises to develop refinement literacy in contracts, refinement, reactivity, and semantic linking. Drawing on experience from university courses, intensive schools, regional programmes, and industrial delivery, we argue that the most effective way to teach UTP is to make procedures explicit, make side conditions visible, and provide fast feedback. We conclude by outlining a lightweight tooling and artefact agenda that supports research-led teaching without imposing the full cost of mechanisation.

Keywords: Unifying Theories of Programming · formal methods education · refinement · semantics

1 Introduction

Teaching formal methods always involves a trade-off. On the one hand, realistic semantic models must account for concurrency, interaction, partiality, time, and hybrid behaviour. On the other hand, students learn only when they can reuse a small number of stable ideas and receive fast feedback on whether a step is

© The Author(s), under exclusive license to Springer Nature Switzerland AG 2026
G. Carvalho and T. Kobayashi (Eds.): FMTea 2026, LNCS 16566, pp. 12–28, 2026.
https://doi.org/10.1007/978-3-032-26743-6_2

correct. In practice, the main teaching difficulty is therefore not that students cannot read the definitions, but that they cannot see what they should do *next*. They need a *method* for progressing: something to appeal to when they get stuck. More broadly, formal methods are part of both the intellectual core of computer-science education and professional practice in industry (see Broy et al. [2] and ter Beek et al. [1]). Computer science education should also cultivate what has been called *formal methods thinking*: the use of formal ideas in practical, lightweight, and accessible ways (see Dongol et al. [6]). Our starting point in this paper is that UTP is especially valuable for teaching because it provides exactly such a method: a disciplined way to specify, calculate, and refine. A small example shows why this matters.

Example 1. A micro-example (why refinement needs extremes).
In the design notation (see Hoare & He [12] and Woodcock & Cavalcanti [13]), the contract $(x = 0) \vdash (x' = 0) \wedge (x' = 1)$ (*precondition $\vdash$ postcondition*) demands an impossible guarantee; in the refinement lattice, it collapses to a *miracle*: something that cannot be satisfied by any executable code, the top of the lattice. Rather than being merely a curiosity, this yields an immediate teaching payoff: students can pinpoint the exact strengthening step that rendered a requirement unimplementable, and the calculus reveals the problem early as a precise diagnosis.

UTP helps this trade-off between reality and pedagogy by separating what students must learn once from what they can add later. At its centre is a common calculus of relations, contracts, and refinement. Around that core, each theory introduces its own observables, healthiness conditions, and linking constructions. Students begin to see the value immediately. We can teach one small calculus well and then reuse it with controlled extensions across multiple paradigms. They first learn to calculate with relations, designs, and refinement; when they later move to reactive models (see Hoare & He [12], Cavalcanti & Woodcock [4], and Foster et al. [9]) or other theory families, the proof habits remain stable even though the observation space and semantic detail become richer.

This view does not make UTP automatically easy to teach. The abstraction is front-loaded: students must grasp alphabet, healthiness, and order-theoretic reasoning before they fully see the payoff. The difficulty becomes more pronounced when students encounter multiple presentations of the same theory. UTP admits denotational, operational, and algebraic accounts, but unless the links among them are made explicit, students can easily confuse plurality with duplication. The teaching task is therefore not just to present several semantics, but to show why they belong together and what each one contributes. This argument leads to the central question of the paper:

Question *How can we teach UTP as a disciplined method for specifying, calculating, and refining programs, rather than as a collection of definitions and notations?*

Our answer is to organise teaching around three explicit unifications, dimensions that students repeatedly traverse. First, we link theories horizontally, showing how different paradigms share a common relational and refinement core. Second, we move vertically through abstraction levels, showing how we connect requirements, contracts, designs, and implementations via refinement, a meaning-preserving, strengthening development relation. Third, we treat semantic presentations themselves as a teaching object, showing how we link the denotational, operational, and algebraic views by explicit soundness, adequacy, and characteristic refinement arguments.

This paper does not propose a new UTP theory. Its contribution is pedagogical: a *method* for teaching UTP, together with reusable teaching artefacts distilled from repeated delivery across university courses, intensive schools, regional programmes, and industry-facing short courses. More specifically, we contribute a course architecture centred on a minimal teachable core and a reusable taxonomy of exercises to develop refinement literacy in contracts, refinement, reactivity, and semantic linking. A secondary theme is that *semantic variation* should be teachable in its own right. Rather than presenting undefinedness, partiality, and related corner cases as isolated exceptions, we argue that we should introduce them as *controlled semantic choices*, with lightweight tooling and mechanised artefacts used to make those choices visible without imposing the full cost of proof engineering.

The remainder of the paper develops this teaching programme. Section 2 identifies the smallest UTP core that students must internalise. Section 3 turns that core into a staged course design, with assessments and a reusable exercise taxonomy. Section 4 discusses the tool support most helpful for teaching, and Sect. 5 develops the corresponding research-led artefact agenda. Section 6 closes with practical lessons and a proposal for shared community resources.

2   UTP for Teaching: What Students Must Internalise

In the Introduction, we argued that we teach UTP best as a disciplined method for specifying, calculating, and refining. This section identifies the smallest conceptual core that students must internalise for that method to work. The three unifications introduced earlier remain the organising view of the subject: UTP links theories horizontally, links abstraction levels vertically through refinement, and links denotational, operational, and algebraic presentations through explicit proof obligations. However, students do not need to master all of this at once. For teaching, the priority is to isolate the smallest stable core that they can learn early and then reuse across later theory families. That teachable core has four ingredients: alphabetised relational calculus, refinement as an order-theoretic discipline, healthiness conditions, and linking constructions such as Galois connections (based on the foundational work in computer science by Cousot & Cousot [5]). Together, these provide the minimum machinery needed to explain why UTP supports calculation across paradigms rather than merely offering another semantic notation.

Alphabetised Relational Calculus. Students must learn to view predicates and relations as descriptions over an explicit set of observations. The alphabet is not bureaucratic decoration: it records the interface of a specification, determines a predicate's view, and controls composition. Alphabets are pedagogically important because many apparent semantic failures are really interface failures. If students internalise the idea of an alphabet early, they acquire a disciplined way to reason about scope, state change, and compositionality.

Refinement and Lattice Structure. Students must internalise refinement not as an isolated definition, but as the ordering that makes development calculational. At a minimum, they need to understand extremal elements, monotonicity, and the role of joins and meets in comparing specifications. We can defer more advanced fixed-point reasoning until iteration, recursion, or reactive models require it. The key pedagogical point is that refinement tells students which transformations preserve meaning and which strengthenings make a requirement unimplementable. At this point, miracles, infeasibility, and undefinedness become conceptually useful rather than mysterious edge cases.

Healthiness Conditions. Students must internalise that a UTP theory is not just an arbitrary collection of predicates, but a disciplined semantic space characterised by closure conditions. Healthiness conditions explain what counts as a meaningful program in a theory and why certain proof obligations simplify once closure is known. For teaching, this is important because healthiness makes well-formedness a reusable reasoning habit. Before proving anything substantial, students learn to ask whether the object they are manipulating is even a valid inhabitant of the theory.

Linking Constructions. Students must also see that UTP is about principled movement between views, not merely reasoning within a single fixed presentation. Galois connections provide a compact and teachable account of this movement. They explain how abstraction and concretisation relate theories, how operators transport across representations, and why the same design ideas recur in apparently different semantic settings. Pedagogically, linking matters because it gives substance to unification: students see not just that two theories are similar, but how and why.

These four ingredients are sufficient to support the paper's main teaching messages. They explain how paradigms can share a common core, how contracts connect to implementations by refinement, and how we relate multiple semantic presentations by explicit proofs rather than treating them as competing definitions. They also justify a staged course design. We teach the relational and refinement core first; we introduce healthiness as a disciplined notion of semantic well-formedness; and we present linking only once students are ready to compare representations or theory families. In this way, we break the abstract power of UTP into a small number of stable ideas that students can repeatedly apply with fast feedback.

More specialised material, including reactive observables, timed and hybrid models, and full fixed-point treatments of iteration and recursion, can then be introduced later as controlled extensions of this core, rather than as prerequisites for understanding UTP in the first place.

3 Course Design: What We Teach and How We Assess It

Section 2 identified the minimal conceptual core that students must internalise if UTP is to function as a reusable method. The next question is how to turn that core into a teachable course structure. Our aim is not to cover every UTP theory, but to design a progression in which students repeatedly practise a small number of stable reasoning habits, receive fast feedback, and then extend those habits to richer theories and representations. This section addresses three questions: what background we assume, how we stage the material, and how we assess procedural competence rather than passive familiarity.

3.1 Pedagogical Assumptions and Prerequisites

UTP can be taught successfully without extensive prior exposure to formal semantics, provided that the course is carefully scaffolded (structuring the course so that students build competence in stages, with each stage supporting the next) and organised around realistic modelling examples, hands-on tool support, and progressive introduction of notation and reasoning techniques, as in Cataño's experience teaching Event-B to software engineering students [3]. What students must already possess is a reliable command of basic discrete mathematics and programming notation: sets, relations, functions, Cartesian products, elementary predicate logic, substitution and renaming, and ordinary programming constructs such as assignment, conditionals, and loops. This prior education is enough to support the relational core, namely composition, substitution, alphabet tracking, and small design calculations. Of course, we have a foundation course to teach the required preparatory knowledge if this is needed.

Some prior exposure facilitates a smoother development, but it is not essential. In particular, students benefit if they have seen at least one of the following at a reading level: Hoare logic and weakest preconditions, transition-rule-based operational semantics, or a process-algebraic notation such as CSP. They also need basic proof habits, especially equational reasoning by rewriting and the discipline of explicitly stating side conditions. UTP is learned most effectively by calculation rather than by ad hoc proof search. We introduce other material only when it becomes useful. Order-theoretic intuition about extremal elements, monotonicity, joins, meets, and fixed points is relevant to miracles, undefinedness, iteration, and recursion, but it need not serve as a heavy prerequisite. Likewise, we can defer concurrency and reactivity until the core contract-and-refinement habits are stable. Mechanisation is optional throughout: it can serve as an instructor demonstration, a lightweight checking aid, or an advanced track for students already comfortable with proof assistants.

For mixed cohorts, a short diagnostic in the first week is worthwhile. In practice, a 20–30 min check on sets, relations, substitution, and basic program laws, together with a one-page notation repair kit, is usually enough to identify who needs early support. The pedagogical point is simple: we should develop symbolic fluency early, because later difficulties with UTP are usually methodological rather than notational.

3.2 Course Architecture: a Staged Progression

The course architecture should mirror the staged core introduced in Sect. 2. We therefore organise teaching as a progression from a small relational calculus to richer theories, preserving proof habits wherever possible.

1. **Relations and predicates.** We begin with relational composition, substitution, alphabet tracking, and the core laws of the relational calculus. This mathematical toolkit gives students their first experience of UTP as calculation over explicit observations, rather than notation to memorise.
2. **Designs and the assertional bridge.** We next introduce pre- and post-specifications as designs, together with their connection to familiar assertional reasoning. At this stage, weakest preconditions, partial versus total correctness, and the relation to Hoare-style reasoning are pedagogically valuable because they show that UTP does not replace earlier methods, but reorganises them inside a common semantic framework.
3. **Refinement as method.** Once students can manipulate designs, refinement becomes the central organising idea. Here we emphasise algebraic laws, monotonicity, compositional development, and the role of extremal elements. At this stage, students move from reading specifications to transforming them, thereby acquiring what we call *refinement literacy.*
4. **Linking and representation change.** Only after the relational and refinement core is secure do we introduce linking constructions such as Galois connections. This ordering matters. If we introduce linking too early, it looks abstract; if we introduce it after students already trust the core calculus, it becomes a compelling explanation of how specifications and laws move between semantic presentations.
5. **Concurrency and reactivity.** Reactive processes, reactive designs, reactive contracts, and optionally, timed models, are presented as controlled extensions of the same discipline. The observation space becomes richer, but the proof habits remain recognisable: identify the interface, state assumptions and guarantees clearly, and reason by refinement.
6. **Case studies.** End-to-end examples repay the abstraction cost. A small number of carefully selected case studies is preferable to broad coverage because it allows students to see that the same core reasoning patterns persist across different semantic settings and abstraction levels.

This sequence separates a *minimal teachable core* from later theory-family material. In a semester course, we can often cover the whole progression. In a short school or intensive programme, the same architecture scales down naturally: the relational core, designs, and refinement still form a coherent course, even if linking and reactivity appear only in demonstration. The central design principle is therefore not coverage, but stability: each new stage should preserve as much of the previous reasoning discipline as possible.

3.3 Learning Outcomes and Assessment Design

Because we intend the course to teach a method, we must state the learning outcomes in operational terms. We assess not whether students can repeat definitions, but whether they can calculate, diagnose, transport, and refine. By the end of the core part of the course, students should be able to demonstrate the following capabilities:

- **Alphabet competence:** compute and track alphabets, and diagnose failures of composition as interface mismatch rather than as mere logical error.
- **Semantic calculation:** calculate the meaning of a small program in a chosen UTP theory, especially at the relational and design levels.
- **Healthiness reasoning:** prove that a predicate or operator is healthy, and use closure to simplify proof obligations.
- **Refinement literacy:** establish refinements, apply algebraic laws correctly, and derive more concrete specifications or code from abstract contracts.
- **Linking competence:** use a linking construction, typically a Galois connection, to transport a specification across representations and explain what is preserved or reflected by the transport.
- **Reactive competence (later material):** re-express a state-based specification as a reactive contract, identifying new observables, and assumption-commitment contracts.

Table 1 maps these learning outcomes to the exercise types that train them and the assessed artefacts through which we evaluate them.

Each outcome is trained through routine exercises and then tested through a small number of assessed artefacts. The underlying principle is alignment: assessment should reinforce the same procedural habits the course aims to teach. In particular, we should assess refinement literacy through multi-step derivations with explicit side conditions, not only through results. Likewise, linking competence should be assessed not merely by whether a transported specification is correct, but by whether students can explain the semantic content of the transport. In this way, assessment remains faithful to the course's central claim: students should learn UTP as a disciplined way of progressing.

3.4 A Taxonomy of Practical Exercises

Experience suggests that students learn UTP best through many small calculations before they tackle larger case studies. The most useful exercise types are the following.

Table 1. Mapping learning outcomes to exercise types and assessed artefacts.

Outcome	Trained by	Assessed via
Alphabet competence	Alphabet drills, interface-mismatch diagnosis	Quiz/worksheet, marked "explain the failure" items
Semantic calculation	Small relational and design calculations	Weekly problem sets, "derive the meaning" questions
Healthiness reasoning	Closure drills, healthiness proof patterns	Short proof exercise, optional mechanised extension
Refinement literacy	Refinement steps, design algebra, law-driven derivations	Coursework derivation, project milestone from contract to code
Linking competence	Transport exercises with abstraction/concretisation pairs	Coursework on representation change, short reflective explanation
Reactive competence	"Phase change" exercises, contract extraction	Project milestone, case-study report section

Alphabet Drills. These train students to compute alphabets, track before/after variables, and diagnose ill-formed compositions as interface mismatches. Their pedagogical value lies in making interfaces explicit. A representative exercise asks whether $P \; ; \; Q$ is well-formed for two given predicates and, if not, requires the student to repair the interface by renaming or extending observables.

Closure Drills. These introduce healthiness as a calculational notion rather than an abstract slogan. Students apply a healthiness function to a concrete predicate, then check idempotence and monotonicity on small instances. The main learning goal is to see that healthiness closure both characterises a theory and reduces later proof obligations.

Design Algebra. These exercises develop fluency with pre- and post-specifications, precondition propagation, and extremal behaviours. A typical task is to simplify a design expression, compute its precondition explicitly, and identify when an inconsistent guarantee collapses to an extremal element under the refinement ordering. This moment is often the first point at which the miracle and the infeasible become pedagogically useful.

Refinement Steps. These are the core exercises for building refinement literacy. Students prove small refinements using named laws, monotonicity, and direct calculation, often comparing two derivations of the same result. The emphasis is on making side conditions explicit, because this is what distinguishes disciplined derivation from symbolic guesswork.

Linking Exercises. These show that semantic transport is itself a form of reasoning. Given an abstraction-concretisation pair, typically organised as a Galois connection, students transport a specification across representations and state what is preserved, reflected, and lost. The aim is not just to practise a definition, but to make unification mathematically concrete.

Reactive "Phase Change" Exercises. We introduce these only after the state-based core is secure. Students begin with a design over ordinary before/after variables and re-express it as a reactive contract with traces, refusals, waiting conditions, or other observables. The pedagogical point is that reactivity should feel like a change of semantic phase, not a completely new subject.

Taken together, these exercise types provide a practical route from symbolic fluency to semantic understanding. They also scale well across delivery formats. Short schools can emphasise alphabet drills, design algebra, and refinement steps; longer courses can add linking and reactive phase changes; industrial delivery can use the same taxonomy in a lighter, demonstration-led form.

The broader lesson is that we should teach UTP through repeated, law-guided practice rather than through rare, large proofs. Once students can perform small calculations reliably, larger developments become much less mysterious.

4   Tool Support and Current Limitations

Tool support matters in teaching UTP because it turns law-guided calculation into fast, objective feedback. More generally, if formal methods are to be teachable and adoptable in practice, they must be applicable in that their supporting artefacts and workflows are usable in their intended context [11]. Students often struggle less with the underlying mathematics than with checking whether a rewrite is legal, identifying the side conditions generated by a step, and understanding why a refinement fails. The pedagogical goal of a teaching tool is therefore not full automation, but transparency: it should help students see which law was applied, what obligations we created, and why a step succeeds or fails.

4.1   What Feedback Students Need

Across courses, schools, and industrial delivery, the same feedback needs recur.

Law Application and Rewriting. Students need immediate confirmation that a rewrite and its side conditions are legal. In UTP, these side conditions are often the real content of the reasoning: healthiness assumptions, alphabet constraints, monotonicity requirements, and definedness obligations.

Alphabet and Interface Discipline. Many apparently mysterious failures in composition or refinement are simply interface errors. A useful tool should therefore explicitly report missing, renamed, or incompatible observables, rather than returning an opaque proof failure. This lesson reinforces one of the paper's core teaching points: alphabets are semantic interfaces, not bureaucratic decoration.

Healthiness and Closure. Students benefit from being able to apply a healthiness operator mechanically and inspect the resulting normal form. This exercise turns closure from a metatheoretic slogan into an operational check, and makes it easier to see why certain proof obligations simplify once healthiness is known.

Refinement Checking and Local Diagnosis. For teaching, it is often enough to check small refinement steps rather than complete developments. Stepwise refinement checking prevents misconceptions from persisting for weeks, especially when the tool indicates whether the failure arises from an overly strong guarantee, an overly strong assumption, or a missing side condition.

Counterexamples. When a refinement does not hold, a small counterexample state or trace is pedagogically transformative. It tells students not only that they are wrong but also how they are wrong. In teaching contexts, even very small model-finding or executable micro-semantics can have disproportionate value.

Reactive Intuition. Reactive models introduce observables such as traces, refusals, and waiting conditions that are hard to understand from equations alone. Small visual aids, for example, trace growth, event timelines, or simple composition diagrams, can materially reduce the common misunderstanding that reactive semantics is just state plus time.

These feedback loops suggest a simple design principle: teaching tools should expose the structure of a derivation rather than hide it. Their role is to make laws, interfaces, and obligations visible to students as they learn to use them.

Table 2. Mapping exercise families to tool support.

Exercise family	Tool support that helps most
Alphabet drills	Alphabet comparison, interface mismatch reporting, renaming support
Closure drills	Healthiness application, normalisation, basic closure checks
Design algebra	Design normalisation, precondition propagation, extremal-element simplification
Refinement steps	Stepwise refinement checking, named-law replay, side condition reporting
Linking exercises	Transport of specifications and laws across an abstraction/concretisation pair; small-instance checking of adjunction obligations
Reactive phase-change exercises	Trace/event visualisation, alphabet growth checks, basic reactive-law rewriting

4.2   A Capability Map from Exercises to Tool Support

The exercise families in Sect. 3 suggest a corresponding map of helpful tool features, as summarised in Table 2. The point is not that every course requires all of them, but that different kinds of exercises benefit from different levels of automation. In practice, instructors can obtain much of the pedagogical benefit without full mechanisation. Rewriting support, explicit side condition accounting, and lightweight refinement checking already address many of the dominant student error classes.

4.3   Four Tooling Tiers for Teaching UTP

We distinguish four tooling tiers, each appropriate to different teaching contexts.

Tier 0: Paper-and-Pencil with Structured Feedback. This approach works when class sizes are small and tutorial support is strong. Its success depends on quick feedback: stable law sheets, worked examples, clear marking rubrics, and rapid correction cycles. Tier 0 supports the relational and design core well, but it is less reliable for reactive topics and for cross-presentational derivations.

Tier 1: Rewriting and Side Condition Tracking. A lightweight rewriting environment can deliver substantial pedagogical value. At this tier, the tool needs only to encode a small law base, report side conditions as first-class outputs, and normalise designs or contracts into forms suitable for checking and marking. For many short courses and intensive schools, Tier 1 offers the best balance between benefits and setup costs.

Tier 2: Lightweight Mechanisation. We use a proof assistant in a constrained teaching context: a fixed library of definitions and lemmas, a small tactic vocabulary, and carefully scaffolded exercises (see Foster et al. [8, 10]). One workable style hides most of the alphabet machinery within a fixed-state model, allowing students to focus on relations, designs, and refinement. Another makes alphabet extension explicit, so that interface failures appear as concrete proof obligations. Tier 2 is particularly useful for objective checking of submitted derivations.

Tier 3: Full Mechanised UTP. Full mechanisation supports the strongest teaching and research claims: machine-checked linking, larger case studies, and mechanised relationships among denotational, operational, and algebraic presentations. The drawback is overhead. Installation, environment management, and proof engineering can easily dominate short teaching formats unless the environment is carefully curated.

These tiers are not competitors. They form a progression. A course may primarily teach at Tier 0 or Tier 1, drawing occasional demonstrations from Tier 2 or Tier 3 to show students what a fully checked development looks like.

4.4   Current Limitations in Teaching Contexts

Even where strong mechanisation is in place, several limitations are particularly relevant in the classroom.

Setup Friction. Installation and configuration remain major barriers, particularly for short courses and heterogeneous cohorts. In practice, a single-command setup or containerised environment can determine whether tool use succeeds at all.

Proof-Engineering Overhead. Proof engineering can become a major pedagogical distraction, because students must divide their attention between understanding the semantic concepts and coping with the practical mechanics of interactive proof, including syntax, tactic selection, proof-state interpretation, and library search. Students can easily confuse mastery of the tool with mastery of the semantics. We need to curate teaching libraries aggressively: fewer lemmas, better-named lemmas, and guided proof patterns, rather than open-ended proof search.

Opaque Failures. A failed proof is pedagogically useful only if the reason for failure is intelligible. Missing side conditions, incorrect rewrite directions, and alphabet mismatches should be explicitly brought to the surface rather than buried within generic subgoals.

Weak Counterexample Support. Refinement failures without counterexamples are costly in terms of teaching time. Where possible, executable microsemantics or small-instance model finding should provide the smallest failing example.

Limited Reactive Visualisation. Many proof environments are text-centric. For reactive models, however, even modest visual support for traces, events, and synchronisation can materially improve understanding.

4.5   A Pragmatic Wishlist

A teaching-oriented tooling agenda should remain modest and concrete. The most valuable capabilities are the following.

1. **Curated law libraries with explanations.** Each law should expose its side conditions and state, in one line, what reasoning move it supports.
2. **Explicit side condition accounting.** Alphabet, healthiness, monotonicity, and definedness obligations should be accumulated and displayed as first-class artefacts.
3. **Replayable derivations.** Stepwise calculations should be checkable, annotatable, and replayable, both for marking and for reflection.

4. **Executable micro-semantics for counterexamples.** For very small languages, tools should generate counterexample states or traces when a refinement fails.
5. **Reproducible teaching environments.** Lightweight distribution matters: a stable build command, container image, or browser-based deployment often matters more pedagogically than deeper automation.

The overarching teaching goal is transparency. A useful tool should show not only that a refinement is correct, but also which laws we used, what obligations we created, and which obligations failed when the derivation did not go through. Section 5 turns this pragmatic picture into a broader research-led agenda for reusable artefacts, semantic variation packs, and community benchmarks.

5 A Research-Led Artefact Agenda for Teaching UTP

Section 4 argued that teaching tools are most useful when they make laws, side conditions, interfaces, and failures visible. The next question is which research outputs most improve teaching. We claim that the most valuable artefacts are not simply larger mechanisations but rather reusable teaching modules that expose semantic variation, connect multiple presentations of the same theory, and package those connections in forms that instructors can adopt without incurring the full cost of proof engineering.

5.1 Semantic Variation as a Teaching Objective

A recurring pedagogical difficulty is that students meet undefinedness, partiality, and related phenomena as isolated edge cases: division by zero, nontermination, partial pattern matching, or different readings of conditionals with undefined operands. In fact, these are not accidents but semantic design choices. A research-led teaching strategy should therefore make the logic layer explicit and parametric, so that instructors can present different logics as controlled alternatives rather than as ad hoc exceptions.

Across UTP theories, this variation affects several familiar points: the treatment of undefined expressions, the interpretation of conditionals, and the soundness of refinement and healthiness laws. The pedagogical benefit is immediate. Logic plurality gives a precise answer to the question of why apparently similar laws do not always coincide: they depend on different commitments about evaluation, definedness, and extremal behaviour. Presented in this way, semantic variation discourages memorisation of the law and instead encourages students to ask the right question: under which semantic commitments is this step valid?

The corresponding research opportunity is to package these choices in a form ready for teaching. Useful artefacts include logic-profiled law sheets, small counterexample packs that distinguish the profiles, and replayable micro-examples in which one language fragment is rerun under different logic choices to show exactly what changes and what remains invariant. Such artefacts would turn what we often treat as technical overhead into one of the clearest demonstrations of UTP's unifying power.

5.2 Cross-Presentational Modules and the Semantic Triangle

The paper has argued that denotational, operational, and algebraic presentations form a third axis of unification. A research-led agenda should make this axis practical in teaching by packaging it as a reusable module rather than leaving it as a meta-level claim. For a small language fragment, the aim is not merely to present three descriptions side by side, but to prove how they are connected.

A useful format is the *semantic triangle* module, comprising four ingredients: a denotational definition, an operational presentation, an algebraic law set, and at least one explicit result that connects them, such as a soundness, adequacy, or reconstruction theorem. Even partial instantiations of this pattern have strong teaching value. Students see that operational rules are derived rather than guessed, that algebraic laws arise from proof obligations rather than stylistic preference, and that we justify different presentations by proof rather than by authority.

The most valuable research outputs here are derivation templates and one-theorem modules. A derivation template should show how we generate characteristic obligations from a proposed operational rule and then simplify them into lecture-ready proof sketches. A one-theorem module should package a small but explicit adequacy or reconstruction result, together with clearly stated assumptions such as monotonicity or healthiness closure. These modules would provide instructors with research-backed exemplars of the paper's z-axis, while remaining small enough to teach.

5.3 Reusable Benchmarks and Teaching Kernels

If these ideas are to scale beyond individual courses, we need to package them as reusable artefacts. The most promising route is a small benchmark suite designed explicitly for teaching. Such benchmarks should not aim for industrial realism; they should aim for manageable complexity with a clear pedagogical payoff. Each benchmark should be small enough to teach in a short block, but rich enough to exercise the three unifications identified in the paper.

A suitable benchmark should therefore include at least two related semantic views of the same system, a clear path from requirements to a contract to a design or code fragment, and at least one explicit link among denotational, operational, and algebraic accounts. It should also carry a small set of assessment hooks aligned with Sect. 3: an alphabet or healthiness diagnostic, a refinement derivation, and a linking or transport exercise. In this way, the benchmark becomes not merely an example but a reusable teaching unit.

We should also package research artefacts as *teaching kernels*. A teaching kernel is a deliberately small proof-assistant or formalisation bundle containing only the definitions, law sets, derivation scripts, and optional extensions needed for a given cohort. The key principle is curation. Stable law names, explicit side conditions, replayable derivations, and reproducible environments matter more pedagogically than maximal generality. The value of research-led teaching is therefore not that every student sees the full mechanisation, but that the mechanisation yields compact and reusable course artefacts.

5.4 Evidence, Comparison, and Community Reuse

A research-led agenda also requires better evidence. Teaching papers in formal methods often describe promising approaches but report too little that is comparable across cohorts. For UTP, the most useful measures align with the capabilities identified in Sect. 3: whether students can carry out a multi-step refinement derivation with correct side conditions, diagnose alphabet and healthiness failures accurately, transfer a law-guided method from one theory family to another, and use tool feedback to improve both correctness and explanation.

The aim need not be large-scale educational experimentation. Even small cohorts can contribute useful evidence if instructors report common misconceptions, characteristic student errors, and which forms of feedback reduce them. Over time, this suggests a realistic community outcome: a shared repository of exercise banks, benchmark modules, semantic-triangle exemplars, logic-variation packs, and teaching-oriented tool configurations. Such a repository would provide FMTea [7] and the wider UTP community with a practical means to accumulate teaching knowledge rather than repeatedly rediscovering it on a course-by-course basis. These reusable artefacts, rather than comprehensive mechanisation alone, are the most promising route to a durable teaching practice for UTP.

6 Closing Remarks

We can teach UTP effectively, but only if we present its unifying ideas as a practical method rather than as a collection of definitions. The central claim of this paper has been that students learn UTP best when we teach using stable forms of semantic movement: across theory families, across abstraction levels, and across denotational, operational, and algebraic presentations. This organisation provides them with more than a conceptual overview. It provides them with a repeatable way to progress: calculate, simplify, apply a law, identify the side conditions, and justify the refinement step.

Two practical lessons follow. The first is that we should teach procedures, not just concepts. Students make progress when we build the course around a small relational and design core, repeated law-guided exercises, and carefully staged extensions to linking and reactivity. The second is that we must make side conditions visible and easy to check. Alphabets, healthiness, monotonicity, and definedness obligations are not secondary details; they are the point at which semantic soundness becomes learnable practice. When these obligations remain implicit, students tend either to memorise laws mechanically or to treat UTP as a mere exercise in symbol manipulation. When they are exposed explicitly, they become part of the method itself.

The paper has argued for aligned course design, lightweight, feedback-oriented tooling, and research-led teaching artefacts. The goal is not to mechanise everything for every cohort, but to package the right kinds of support: exercise banks with rubrics, small benchmark case studies, replayable derivations, teaching kernels, logic-variation packs, and semantic-triangle modules that relate denotational, operational, and algebraic views. Such artefacts enable teaching

UTP with fast feedback and controlled semantic richness, without incurring the full overhead of proof engineering.

The broader opportunity is community reuse. If FMTea and the UTP community aim to improve teaching practice, the most useful next step is not simply to produce more slides or more mechanisations in isolation, but to share compact, comparable, and reusable teaching modules. A common repository of examples, exercise sequences, logic profiles, semantic-triangle exemplars, and lightweight tool configurations would enable instructors to accumulate pedagogical knowledge collectively rather than rebuild it course by course. That, in turn, would make UTP easier to teach not by simplifying the theory itself, but by making its methods more visible, more repeatable, and more widely shared.

Acknowledgements. I thank the colleagues who helped me to teach UTP over the last 28 years: in particular, Tony Hoare, Ana Cavalcanti, and Simon Foster. I thank many other lecturers and the students who attended the courses. I hope you had as much fun as I did.

References

1. ter Beek, M.H., et al.: Formal methods in industry. Formal Aspects Comput. **37**(1), 7:1–7:38 (2025). https://doi.org/10.1145/3689374
2. Broy, M., et al.: Does every computer scientist need to know formal methods? Formal Aspects Comput. **37**(1), 6:1–6:17 (2025). https://doi.org/10.1145/3670795
3. Cataño, N.: Teaching formal methods: lessons learnt from using Event-B. In: Dongol, B., Petre, L., Smith, G. (eds.) FMTea 2019. LNCS, vol. 11758, pp. 212–227. Springer, Cham (2019). https://doi.org/10.1007/978-3-030-32441-4_14
4. Cavalcanti, A., Woodcock, J.: A tutorial introduction to CSP in *unifying theories of programming*. In: Cavalcanti, A., Sampaio, A., Woodcock, J. (eds.) PSSE 2004. LNCS, vol. 3167, pp. 220–268. Springer, Heidelberg (2006). https://doi.org/10.1007/11889229_6
5. Cousot, P., Cousot, R.: Abstract interpretation: a unified lattice model for static analysis of programs by construction or approximation of fixpoints. In: Graham, R.M., Harrison, M.A., Sethi, R. (eds.) Conference Record of the Fourth ACM Symposium on Principles of Programming Languages, Los Angeles, California, USA, January 1977, pp. 238–252. ACM (1977). https://doi.org/10.1145/512950.512973
6. Dongol, B., et al.: On formal methods thinking in computer science education. Formal Aspects Comput. **37**(1), 8:1–8:23 (2025). https://doi.org/10.1145/3670419
7. Dongol, B., Petre, L., Smith, G. (eds.): FMTea 2019. LNCS, vol. 11758. Springer, Cham (2019). https://doi.org/10.1007/978-3-030-32441-4
8. Foster, S., Baxter, J., Cavalcanti, A., Woodcock, J., Zeyda, F.: Unifying semantic foundations for automated verification tools in Isabelle/UTP. Sci. Comput. Program. **197**, 102510 (2020). https://doi.org/10.1016/J.SCICO.2020.102510
9. Foster, S., Cavalcanti, A., Canham, S., Woodcock, J., Zeyda, F.: Unifying theories of reactive design contracts. Theor. Comput. Sci. **802**, 105–140 (2020). https://doi.org/10.1016/J.TCS.2019.09.017

10. Foster, S., Zeyda, F., Woodcock, J.: Isabelle/UTP: a mechanised theory engineering framework. In: Naumann, D. (ed.) UTP 2014. LNCS, vol. 8963, pp. 21–41. Springer, Cham (2015). https://doi.org/10.1007/978-3-319-14806-9_2
11. Gleirscher, M., van de Pol, J., Woodcock, J.: A manifesto for applicable formal methods. Softw. Syst. Model. **22**(6), 1737–1749 (2023). https://doi.org/10.1007/S10270-023-01124-2
12. Hoare, C.A.R., Jifeng, H.: Unifying Theories of Programming. Series in Computer Science. Prentice Hall (1998)
13. Woodcock, J., Cavalcanti, A.: A tutorial introduction to designs in unifying theories of programming. In: Boiten, E.A., Derrick, J., Smith, G. (eds.) IFM 2004. LNCS, vol. 2999, pp. 40–66. Springer, Heidelberg (2004). https://doi.org/10.1007/978-3-540-24756-2_4

Assessment and Evaluation in FM Education

Autograding Weakest Precondition
Proofs and Dafny Specifications

Graeme Smith[(✉)][ID] and Hunter Whitlock

School of Electrical Engineering and Computer Science, The University of
Queensland, Brisbane, Australia
g.smith1@uq.edu.au

Abstract. Autograding has been widely used in programming courses
for decades. It reduces the marking load on teaching staff, providing
them with more time to give higher quality feedback. There is relatively
little work, however, on autograding in formal methods courses where
the emphasis is not on programming, but on writing proofs and specifications. This paper describes a tool which employs the proof capabilities
of Dafny to automatically grade students' weakest precondition proofs
and Dafny method specifications. The tool has been designed to (i) be
reusable for new versions of assignments, with minimal input from teaching staff, (ii) allow fine-grained marking, in order to give students part
marks for partially correct solutions, and (iii) not artificially constrain
students' use of programming techniques where they need to provide
code as well as specifications, allowing them to apply best programming
practices. The paper evaluates the use of the tool over two assignments
of a course with 217 undergraduate and Masters students and highlights
lessons learnt.

1 Introduction

Auto-active program verifiers such as KeY [1], Why3 [9] and Dafny [12,13]
have found widespread usage in formal methods courses worldwide [3,4,7,8,17].
They enable students to verify larger, more realistic programs than can rea-
sonably be done with traditional pen-and-paper proofs. For students to engage
successfully with these tools, however, requires that they have some understanding of the underlying foundations on which the tools are built: when the tools
fail to verify a program *auto*matically, then the students need to inter*actively*
provide additional annotations, such as further invariants, to their code. Deduc-
ing what these annotations should be requires an understanding of the theory
behind program verification.

For this reason, *most* such courses start with an introduction to underlying
theory, Hoare logic [11] and/or weakest precondition reasoning [5,6], before students start using a tool in earnest.[1] Hence, there is often assessment involving
proofs by hand. Even for assessment involving the use of tools, marks will be

---

[1] A notable exception is the Victoria University of Wellington course described in [17].

G. Carvalho and T. Kobayashi (Eds.): FMTea 2026, LNCS 16566, pp. 31–49, 2026.
https://doi.org/10.1007/978-3-032-26743-6_3

allocated for correct specifications written in predicate logic. Marking proofs and specifications by hand is time-consuming at best, and tedious and error-prone at worst. This becomes problematic as class sizes grow disproportionately to the number of teaching assistants available for marking. Time allocated per student submission is severely restricted and the quality of feedback to students drops.

This situation arose in a University of Queensland course CSSE3100/7100[2] *Reasoning about Programs* as student numbers increased three-fold from 73 students in 2018 to 217 students in 2025. The course is a first introduction to formal methods offered as an elective to 3rd-year undergraduates and course-work Masters students. These students come with a background in programming, including object orientation, and some exposure (usually just one course) to discrete maths. The course covers both weakest precondition reasoning and reasoning about imperative and object-based programs using the Dafny verifier, largely following Parts 0 and 2 of the textbook *Program Proofs* by Rustan Leino [13].

To alleviate the marking load, during the 2025 running of the course we built and deployed an autograder for all three of the course's assignments: Assignment 1 that tests the students' ability to write weakest precondition proofs, Assignment 2 that focuses on specification and implementation of Dafny methods, and Assignment 3 that focuses on data abstraction using Dafny classes. Due to space limitations, we will report on only the first two assignments in this paper.

The autograder utilises the in-built proof capabilities of Dafny to enable fine-grained checks of students' submissions, allowing part marks to be allocated to partially correct submissions. It is also designed to be reusable for similar assignments in future years, and to not artificially constrain students' use of programming style in assignments where writing code is required. A further decision was that the autograder should not affect the students' experience of the course which had been receiving high student evaluation results since 2018. To this end, we decided that the autograder should not change the established marking criteria, nor should it provide additional help to students before submission. It would only be used after submission, during marking, to automate those parts of the marking which could be mechanised: checking correctness of proof steps in weakest precondition proofs, and adequacy of specifications. The teaching assistants would then provide additional feedback (as they had done in the past) and check non-mechanisable parts of the marking.

The use of autograders in programming courses is well established. In some cases, program verifiers have been used to build such autograders. For example, Milovančević et al. [16] use the program verifier Stainless [10] for autograding functional programs written in Scala. Literature on autograding in formal methods courses, particularly autograding proofs and specifications, is however difficult to find. The *Software Foundations* book series by Benjamin C. Pierce et al. [18] uses an autograder to check students' proofs in Rocq [2] by comparing students' solutions with reference solutions. However, to the best of our

---

2 The course is dual-coded as it is offered to both undergraduate and Masters students.

knowledge its details have not been published. Similarly, Noble et al. [17] mention using the Dafny verifier to autograde Dafny specifications with respect to "oracle" specifications, but do not provide the details in their paper of how this is done.

In this paper, we describe the way our autograder uses Dafny to perform fine-grained checking of student assignments. We hope it can provide guidance for others who may wish to build similar tools. We start in Sect. 2 with a brief overview of Dafny and its proof capabilities utilised in our tool. We then provide an overview of our tool in Sect. 3 and describe how Dafny is used to autograde students' weakest precondition proofs in Sect. 4, and method specifications (and loop invariants) in Sect. 5. Since we developed the autograder while the course was running, occasional wrong decisions were made and we report on these as lessons learnt. An evaluation of our 2025 trial of the tool is given in Sect. 6 before we conclude in Sect. 7.

2 Proofs in Dafny

Dafny [12, 13] supports verification of programs written in functional, imperative and object-based styles (supporting *traits*, similar to Java interfaces, but not full inheritance). In CSSE3100/7100, the focus is on imperative and object-based programs since these are the styles familiar to all students in our cohort.

Standard use of Dafny involves annotating a program with pre- and post-conditions (for methods and functions), and loop invariants and variants (for loops and recursion). Further annotation with assertions and assumptions can help with debugging when verification does not succeed. All annotations, apart from variants, are written in first-order predicate logic.

More advanced use of Dafny can make use of lemmas and stepwise proofs referred to as *proof calculations* [15]. These features allow a more interactive style of verification when the verifier cannot automatically discharge a condition. Lemmas can be called from a method's code when needed (much like a method is called) and their proofs are written either inductively (using recursion) or as a proof calculation. Such proof calculations can also be in-lined directly in code, removing the need for introducing a lemma that will only be called once.

In-lined proof calculations form the basis of our autograding approach. As an example of their use in Dafny, consider the following method M for calculating the minimum number of assignments allocated to a teaching assistant. In the example, the number of students s and teaching assistants t are provided as inputs and the minimum number of assignments allocated to a teaching assistant r is calculated as s/t. All variables are of type nat (natural numbers).

```
method M(s: nat, t: nat) returns (r: nat)
        requires s > 0 && t > 1
        ensures r < s
{
        r := s/t;
}
```

Given the precondition $s > 0$ and $t > 1$, we would expect the postcondition $r < s$ to be true. However, Dafny[3] is unable to verify this due to non-linear arithmetic in the weakest precondition of the method, $s/t < s$. Below we augment the method with an in-lined proof calculation which proves in a stepwise fashion that $s/t < s$ is equivalent to $1/t < 1$, a predicate without non-linear arithmetic which Dafny can verify to be true. With the added proof calculation, the postcondition is automatically verified.

```
method M(s: nat, t: nat) returns (r: nat)
        requires s > 0 && t > 1
        ensures r < s
{

    calc {
        s/t < s;
    == (s/s)/t < s/s;
    == 1/t < 1;
    }
    r := s/t;

}
```

In this example, the lines of the proof calculation are related by == (equivalence, which can also be written as <==>). In general, proof calculation lines can be related by any transitive operator, e.g., ==> (implication).

3   Autograder Overview

To facilitate autograding of assignments, our tool requires students to submit[4] a Dafny file which extends a template file provided to them with the assignment description. The template ensures, for example, that students do not (inadvertently) rename methods or method parameters that the autograder is expecting to find. On submission, the file is checked to be of the right kind (i.e., a .dfy file). It is then checked to be syntactically correct and to conform to the template file (i.e., no lines have been removed or altered). Finally, it is checked for soundness issues. These may occur if the student has used assumptions in their code, included methods with specifications but no code, or included lemmas with a specification but no proof. If any of the submission checks fail, the autograder stops and the student is informed so they can fix the problem and resubmit (see Fig. 1).

The syntax check is performed by calling the command dafny verify with appropriate options set to ensure a reasonable time-out and allow use of deprecated syntax, and the soundness check by calling the command dafny audit,

---

[3] Dafny version 4.11.0.0 within VS Code extension 3.5.2.

[4] We deploy our autograder in the Gradescope online assessment tool: https://www.gradescope.com/.

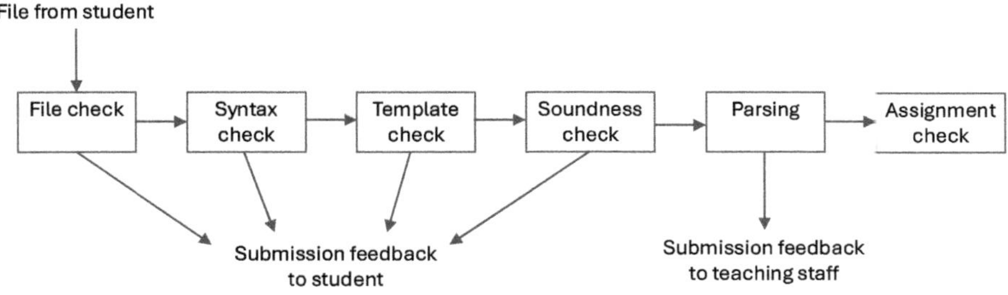

Fig. 1. Autograder design.

again with an appropriate option set to allow use of deprecated syntax. Using dafny verify for the syntax check has two useful side-effects: (i) the verification result can be stored in case needed later for marking purposes, and (ii) by using the –print option of dafny verify a formatted version of the student's file can be created. The formatting ensures that each opening and closing brace appears on its own line (except for the opening brace of a class or an if- or else-statement), indents are uniformly two spaces, all comments (inline and block) are removed and all assignment statements are condensed to one line.

The formatted file is readily parsed by a light-weight parser: one that assumes the formatting, and additionally treats predicates, expressions, and most lines of code as terminal symbols that are not parsed further. The parser is independent of any given assignment in order to be used for all assignments both in the current run of the course, as well as in future years. If the parser fails, this indicates an error in our tool and is reported to the teaching staff to fix (rather than to the student).

4 Weakest Precondition Proofs

Assignment 1 in CSSE3100/7100 requires that students provide a weakest precondition (wp) proof of one or more methods. Generally, as a warm-up, and to provide useful properties for simplifying the wp proofs, they also prove one or more equivalences in predicate logic. Each method usually has at least one loop and one if-statement and, together with supporting functions (calls to which are used in its specifications and invariants), comprises around 20 to 25 logical lines of Dafny code (LLOC).[5] For example, in 2025 there was one method: the third version of the year-calculation algorithm at http://bit-player.org/2009/the-zune-bug which, together with supporting functions, was 22 LLOC. The students had to provide two wp proofs over the method: one to prove partial correctness and one to prove termination. They also had to provide one predicate logic proof.

---

[5] Excluding lines with just opening or closing braces, but including predicate annotations (pre- and postconditions, invariants, etc.).

4.1 Template File

For this assignment, the template file includes the Dafny method(s) and we ask the students to fill in the lines of the proofs between the lines of code. While these proof lines could be added as comments, we want them to be syntax- and type-checked Dafny predicates for two reasons. Firstly, this allows students to detect simple syntax errors, missing brackets, etc. in their predicates and fix them before submission. Secondly, it allows us to use these predicates directly in our tool.

A number of approaches were considered for how students could include predicates within the code block of a method. One possibility was to change the syntax used by the Dafny verifier to include a special keyword for proof steps. This, however, would limit students to using a particular (custom) version of the verifier for this assignment, which did not seem like a good long-term solution. We decided, therefore, to utilise existing Dafny syntax for the predicates. One possibility was to use assertions. The verifier indicates, however, whether or not assertions hold allowing students to detect errors in their proofs (when we would like them to be able to construct such a proof without mechanised assistance). The approach might also cause students to unnecessarily worry about proof steps that are correct but not readily verified by the verifier.

Other annotations that were considered were assumptions and **expect** clauses: the latter are run-time assertions that behave similar to an assumption at verification time. Both were also rejected due to the potential to mislead students as to the use of these Dafny constructs (especially assumptions which appear later in the course) and due to the possibility of unforeseen side-effects. In the end, we adopted a simple approach where the predicates are assigned to special Boolean variables introduced as part of the template: WP (for weakest precondition) used for all predicates except where the students have strengthened the expected predicate (e.g., in order to reach the invariant required at the top of a loop body) and WP_s for predicates when strengthening has occurred.

For example, the wp proof

$$\begin{array}{ll} \{s = n * (n-1)/2 \wedge n! = 33\} & \text{(strengthening)} \\ \{s = n * (n-1)/2\} & \text{(arithmetic)} \\ \{s + n = n * (n-1)/2 + n\} & \\ \mathsf{s} := \mathsf{s} + \mathsf{n}; & \\ \{s = n * (n-1)/2 + n\} & \end{array}$$

is written by the students as the syntax and type-checkable Dafny code below.

```
WP_s := s == n*(n - 1)/2 && n != 33;
WP := s == n*(n - 1)/2;                    // arithmetic
WP := s + n == n*(n - 1)/2 + n;
s := s + n;
WP := s == n*(n - 1)/2 + n;
```

Note that students are required to justify their proof steps with comments (e.g., arithmetic above) and this justification is marked manually by teaching staff. This prevents students performing leaps of logic, which our autograder may judge correct, without understanding why the predicate transformation holds.

For autograding the predicate logic proofs, we employ a similar approach: for each equivalence the students need to prove, we introduce a method with a single line of code introducing a Boolean variable PS (for proof step) and the students provide the proof via assignments to PS.

4.2   Checking Proofs

To check the proofs, for each proof step we create a Dafny program which verifies when that proof step is correct. These programs share a common structure: a class Checker with each of the method's input, output and local variables as fields and a single method for performing the check.

```
...  // supporting functions

class Checker<T(==)(0)> {
        var ν   // input, output and local variables

        ...  // checking method goes here
}
```

The generic parameter T of the class allows for the method being proved to have a generic parameter (we assume one generic parameter is enough for our assignments). The type annotation (==) allows equality to be used to compare elements of T, and (0) allows elements of T to be used without being explicitly initialised, e.g., a non-zero sized array of elements of T may be declared. These are included to allow for common uses of generic types.

Working backwards through a wp proof, the first thing we check is that the final predicate ϕ of the proof is equivalent to the method's postcondition ψ. We do this under assumptions α provided by the teaching staff. These assumptions are assignment-specific and are intended to avoid the verifier failing to verify a correct step. For example, with method M of Sect. 2 we would include the assumption assume{: axiom}s/t < t.[6] The general form of the checking method is as follows.

```
method Check() {
        assume {:axiom} α;
        calc { φ; == ψ; }
}
```

---

[6] The attribute {:axiom} suppresses a warning Dafny would otherwise give when an assumption is used. It is optional since it will not change the outcome of the check.

Assumptions are used rather than assertions, which would have a similar effect, due to it being known that too many assertions can overwhelm the verifier (which needs to prove them all) leading to time-out. We have no experience of the same time-out behaviour occurring with assumptions.

A checking method of the same form is used to check the predicates associated with while loops and if-statements. For example, for an if-statement we check that the predicate at the bottom of the if- and else-branches are equivalent to the predicate below the if-statement, and that the predicate above the if-statement is equivalent to the predicate $(b ==> P)$ && $(!b ==> Q)$ where b is the if-statement's guard, P is the predicate at the top of the if-branch and Q the predicate at the top of the else-branch.

To have access to the required predicates, such as P and Q above, as we move backwards through the proof we store them in stacks. Using stacks allows us to handle nested statements, e.g., in our 2025 assignment we had an if-statement nested in the if-branch of another if-statement which itself was nested in a while loop.

The above checking method is also used to compare the predicates whenever we have an assignment to WP occurring above another assignment to WP or, in our predicate logic proofs, for any consecutive assignments to PS. When we have an assignment to WP_s, however, we need to check that the predicate has been strengthened (and is not equivalent to that below it). This is done by first checking the new predicate ϕ implies the predicate below it ψ

```
method Check() {
        assume {:axiom} α;
        calc { φ; ==> ψ; }
}
```

and then checking that ψ does not imply ϕ (i.e., they are not equivalent). This is done by showing the following checking method fails to verify.

```
method Check() {
        assume {:axiom} α;
        calc { ψ; ==> φ; }
}
```

Our experience with complex predicates ψ and ϕ is that Dafny is more capable of showing the failure of the above implication than showing directly that the predicates are not equivalent.

For transformations of predicates over an assignment x := E, we use the following checking method where ϕ is the predicate before the line x := E and ψ is the predicate after it.[7] The annotation modifies this allows x := E to change any fields of the class Checker. The function old refers to the value of its enclosed expression (in this case the predicate ϕ) at the start of the method.

---

[7] The same method is used for assignments to local variables that occur when the variable is declared, i.e., assignments of the form var x := E.

```
method Check()
      modifies this
{
      assume {:axiom} α;
      x := E;
      assert ψ == old(φ);
}
```

To see why this check works, consider the following wp proof of the method which shows that when the assertion holds (and hence the method verifies), $wp(x := E, \psi)$, i.e., $\psi[x \backslash E]$, is equivalent to ϕ.

$$\{\psi[x \backslash E] \iff \phi\} \qquad \text{(since we are at the start of the method)}$$

$$\{\psi[x \backslash E] \iff \text{old}(\phi)\} \qquad \text{(since old values are constants and not updated)}$$

```
x := E;
```

$$\{\psi \iff \text{old}(\phi)\} \qquad (wp(\text{assert } P, Q) = Q \wedge P)$$

```
assert ψ == old(φ);
```

$$\{true\} \qquad \text{(since there is no explicit postcondition)}$$

Finally, when we reach the top of the proof, we check whether the top-most predicate ψ is implied by the method's precondition ϕ. If this is the case, the proof shows that the method is correct; otherwise, it shows that the method is incorrect. The result of this check is for the teaching staff only (and not shown to students). It allows the staff to quickly check whether the student's proof matches what they claim about it. The required checking method is of the form

```
method Check() {
      assume {:axiom} α;
      calc { φ; ==> ψ; }
}
```

4.3   Discussion

When trialed in the 2025 run of CSSE3100/7100, the autograder worked as expected for Assignment 1 apart from one implementation error (see Sect. 6) and one issue with a supporting function which was partial. To illustrate the latter issue, consider the following function Div, which returns num/den provided parameter den is non-zero.

```
function Div(num: int, den:int): int
      requires den != 0
{num/den}
```

Whenever an application of the function appears within a proof, there should be a conjunct ensuring the application is defined. For example, the proof might have a predicate d > 0 && x == Div(n, d). Students, however, might write this predicate as x == Div(n, d) && d > 0. Although this is logically equivalent (and hence correct in the context of a proof), Dafny will not verify the associated check(s) since (like other programming languages) it evaluates predicates from left to right and hence regards the function application to be potentially undefined.

After discovering several student submissions with this problem, we changed the partial function, which was provided as part of the input to the autograder, to be total, returning an irrelevant constant when called out of its original precondition. For example, we could change Div above to be

```
function Div(num: int, den:int): int {
    if den != 0 then num/den else 42
}
```

Either ordering of the conjuncts above is then acceptable. This approach conforms with our decision not to change the students' experience of the course: in previous years, proofs were checked by the teaching assistants who did not require the conjuncts to be ordered. In any case, taking marks off for wrongly ordered conjuncts would be harsh in an assignment asking for a weakest precondition proof.

Below are some reflections on the autograder design.

Generality. The approach detailed above covers methods with while loops, if-statements and assignments, the kinds of instructions expected to occur in Assignment 1 of CSSE3100/7100. Only method calls are not supported in the current autograder implementation. Initial attempts at adding a checking method for method calls, similar to that for assignment, highlighted the need for additional code in the checking class to ensure that the quantification that arises in the precondition of a method call has adequate triggers [14]. Deciding on the best way to do this has been left for future work.

The approach can be used to check proofs of both partial and total correctness. For the latter, we follow the convention in the course textbook [13] to prove termination. A local variable d0 is used to record the value of the variant at the top of the loop body. The autograder checks whether d0 exists as a local variable and adjusts the expected predicate at the bottom of the loop body to ensure the variant decreases appropriately (see page 248 of [13] for details). Hence, d0 is reserved for this purpose in our assignments.

Checking Proof Steps. One of our design criteria for the autograder is that students should be given partial marks for correct parts of solutions. By considering each proof step in a separate check, we avoid an incorrect step causing later checks to fail; the later checks are done based on the incorrect student predicate, not the predicate that was expected.

To ensure student proofs do not make large leaps which Dafny may verify as correct, but the student themselves might not understand (e.g., when trying to reach a loop invariant in the proof at the top of a loop body), students are required to justify predicate simplifications. Justifications are provided as unconstrained comments and are marked manually by the teaching staff.

Need for the Checker Class. The embedding of the checking methods within a class allows us to use the old function for checking predicate transformations over assignments: old only applies to values on the heap, i.e., arrays and fields of objects. Alternatively, we could check directly that $\psi[x\backslash E]$ (the wp of the assignment) is equivalent to ϕ (the student's suggested precondition), but this would require finer-grained parsing of Dafny predicates than we have currently implemented to enable substitution of free occurrences of x in ψ with E.

5   Method Specifications (and Loop Invariants)

Assignment 2 in our course requires students to specify and implement one or more Dafny methods. The methods will generally involve loops, possibly nested, to test the students' ability to write loop invariants (something that was provided to students in Assignment 1). For example, in 2025 the assignment asked the students to specify and implement Cycle Sort https://en.wikipedia.org/wiki/Cycle_sort and a second method which uses Cycle Sort to efficiently find the smallest positive number that is not in an array. While coding is required in this assignment, it is assumed that this is not difficult for the students and the emphasis, in terms of marking, is on their ability to write pre- and postcondition specifications and loop invariants.

5.1   Template File

The template file, in this case, is minimal – just the headers of the methods that the students must specify and implement. This ensures the autograder can find these methods, i.e., the students have not renamed them nor their parameters. Since one of our design goals is allowing the students flexibility in their coding, we allow them to add helper methods to the template file. Also for flexibility, we allow them to add functions, which they can use to simplify their specifications (and invariants), and lemmas, which they can use, when required, to assist Dafny with verification.

5.2   Checking Specifications

For the pre- and postconditions, the autograder checks the students' specifications against a solution provided as input by the teaching staff. Often these specifications will comprise multiple conjuncts and we check that (i) the full student pre- or postcondition implies each of the conjuncts in the solution (and hence is not too weak) and (ii) the full solution pre- or postcondition implies each of the conjuncts of the student (and hence these are not too strong).

To enable (i), we write each conjunct in a separate **requires** or **ensures** clause in our solution. This allows the conjuncts to be readily extracted using our lightweight parser (which does not parse predicates). Marks can be given for correct conjuncts provided by the student even if other expected conjuncts are missing.

It is not possible to similarly extract the conjuncts of the students specifications (where there may be more than one per **requires** or **ensures** clause). However, this is not a problem as we have a fixed penalty for a pre- or postcondition being too strong. The failure of the full solution pre- or postcondition to imply the predicate of any student **requires** or **ensures** clause is sufficient for that penalty to be applied.

Additionally, we check that the student has not (inadvertently) written conflicting conjuncts which makes their full pre- or postcondition equivalent to false. If this is the case, they do not get any marks for (i) and (ii).

As in the first assignment, we carry out each check within a class, Checker. In this case, as well as supporting functions provided as part of the assignment, we include in our program any functions additionally provided by the student.

```
...  // supporting functions (from assignment)
...  // additional functions (from student)

class Checker<T(==)(0)> {
    ... // checking methods go here
}
```

Preconditions. The first check carried out is that a student's full precondition Ψ is not false (under any assumptions α provided by the teaching staff). The check for this is shown below where ν_{in} is the list of the method's input declarations.

```
method Check(ν_in) {
    assume {:axiom} α;
    calc { Ψ; ==> false; }
}
```

When the above check fails (i.e., Ψ does not imply false), the student's full precondition is checked to be strong enough by showing that it implies each conjunct of the solution. The checking method for one such conjunct ϕ_i is as above with the proof calculation replaced with calc { Ψ; ==> ϕ_i; }.

To show the student's full precondition is not too strong, we show that the full solution precondition Φ implies each of the **requires** clauses ψ_i of the student. The checking method is as above with the proof calculation replaced this time with calc { Φ; ==> ψ_i; }.

Postconditions. In the case of postconditions, the students' predicates might contain old values. Hence, we need to perform our check over two states. To create a post-state distinct from the pre-state, we introduce a method Change which arbitrarily updates any variables in the method's modifies list; this list only includes heap-based variables which in Assignment 2 will be arrays.

When checking whether a student's full postcondition Ψ implies false, for example, the checking method first calls Change to create a new state. This new state is the 'post-state' in which we check the student's postcondition and the initial state (before Change is called) is the 'pre-state'. Since postconditions might only be defined when the pre-state values they refer to satisfy the precondition, we add an assumption that the precondition pre of the method being checked is true in the pre-state. The checking methods are shown below where ν is the list of both input and output declarations, μ_{in} the list of input variables, and δ the list of variables in the method's modifies list.

```
method Change($\nu_{in}$)
     modifies $\delta$

method Check($\nu$)
     modifies $\delta$
{
     assume {:axiom} $pre$;
     Change($\mu_{in}$);
     assume {:axiom} $\alpha$;
     calc { $\Psi$; ==> false; }
}
```

As for preconditions, when the above check fails, the student's full postcondition is checked to be strong enough by showing that it implies each conjunct of the solution. The checking methods for one such conjunct ϕ_i are as above with the proof calculation replaced with calc { Ψ; ==> ϕ_i; }.

To show the student's full postcondition is not too strong, we show that the full solution postcondition Φ implies each of the **ensures** clauses ψ_i of the student. The checking methods are as above with proof calculation calc { Φ; ==> ψ_i; }.

5.3 Checking Invariants

Initially, we believed that by requiring the students to write their code using a particular structure (in terms of loops and loop nesting for each method) that we could check their loop invariants using the above approach for pre- and postconditions. Loop invariants need to be strong enough to imply the method's postconditions and may be stronger provided they are not so strong that the method fails to verify, e.g., by not being true when the loop is first reached. Given that we planned to do a verification check of each method, we did not need to check whether the invariants were too strong and hence focused on just checking that they were not too weak.

Although this worked for the majority of students, two problems arose. Firstly, students who had the wrong postcondition and therefore had invariants different to our solution felt they were marked too harshly, losing marks for invariants which were correct with respect to their (incorrect) postconditions. Secondly, we discovered during the marking phase that a number of students had used local variables in their code and these appeared in their invariants. Hence, their potentially correct invariants were not able to be compared with the solution.

The second problem is one that is not easily remedied given that we do not want to constrain students' programming style. We marked the affected assignments' invariants manually, but have since developed an approach that solves both problems. Rather than checking the invariants directly, for each correct conjunct of a student's postcondition we check that the postcondition is ensured by the student's implementation of the method. For this to succeed, the parts of the student's invariants that ensure that conjunct must be correct.

The checking method is below where ν_{out} is the list of output declarations, ψ_{ens1} to ψ_{ensi} the first i conjuncts of the student's postcondition and $code$ the student's code including loop invariants. When checking ψ_{ensi} is ensured by the code, we include each of the previous conjuncts which has passed, i.e., which is ensured by the code. This is necessary due to the fact that conjuncts are evaluated in order in Dafny and hence later conjuncts might depend on earlier ones to be well defined.

```
method Check(ν_in) returns (ν_out)
      requires pre
      modifies δ
      ensures ψ_ens1 // only when passed
         ...
      ensures ψ_ensi
  {
      code;
  }
```

Note that we do not add assumptions in this check. The students are required to use assertions or lemmas, if necessary, to get their code to verify.

Since the method being checked could have calls to other methods of the assignment, the specifications of these are included in the overall program. We include the solution's method specifications, not the student's, to avoid errors in one student method affecting the checking of other methods in which that method is called. Also, additional lemmas or methods provided by the student, e.g., helper methods used to simplify their code, are included in full.[8]

---

[8] Only additional lemmas and methods which verify are included. Since they are included in all invariant-related checks, failing lemmas or methods would otherwise cause even checks which do not call them to fail.

5.4   Discussion

For this assignment, the light-weight parser is again used as we do not need to parse predicates. The solution's **requires** and **ensures** clause each have a single conjunct of the overall pre- or postcondition, and hence these conjuncts are available. The individual conjuncts of the students' submissions are not required as a fixed mark is deducted when a student's pre- or postcondition is too strong.

The class **Checker** is not strictly required in this case, but is used to reuse the infrastructure developed for the first assignment.

Apart from the issue already discussed (for checking invariants), the autograder ran as expected. Manual marking was used for checking that students had provided adequate justification for their invariants. This was required to encourage students to systematically derive invariants from postconditions (following the approaches in [13]), rather than find invariants by a process of trial and error.

Developing a sufficient set of assumption for this assignment was less straightforward given the variety of student specifications. In 2025, the postcondition of Cycle Sort required the use of multisets to capture that the elements in the sorted list are unchanged. Assumptions were needed to allow reasoning with multisets and these varied depending on the form of the postcondition. To cater for all forms that students used, we adapted the set of assumptions (i) before submission, when students posted questions about their solutions privately on the course discussion forum, and (ii) after submission, when anomalies were detected during manual marking.

6   Evaluation

The decision to not change the marking criteria of assignments resulted in grade distributions comparable to those of previous years. In terms of evaluating the autograder, we had two main objectives. Firstly, to find out by how much marking time is reduced, and secondly to find out about any increase in accuracy of marking. The mean marking times in minutes per submission for 2024 and 2025 are shown in Table 1.

Table 1. Mean marking times in minutes.

Year	Assignment 1	Assignment 2
2024	20.08	21.92
2025	9.87	5.05

For Assignment 1, the average time per submission dropped from 20.08 min in 2024 to 9.87 min in 2025, representing approximately a 50% reduction. Assignment 2 saw an even more substantial improvement, with average marking time decreasing from 21.92 min to 5.05 min, approximately a 77% reduction.

The difference in improvements between the two assignments reflects the manual marking that was required. In Assignment 1, justification of proof steps was checked. This involved understanding the individual proofs steps and checking whether the justification provided for them made sense, a time-consuming task. Although justification of loop invariants was required for Assignment 2, there was much less variability between submissions as only a small number of justifications could be used.

To remove any bias introduced by having different teaching staff in 2024 and 2025, we additionally compared the marking times for three staff members who were involved in marking in both years (see Table 2). Among this group, the average time to mark Assignment 1 fell from 16.92 to 11.88 min (approximately a 30% reduction), and Assignment 2 from 17.89 to 6.41 min (approximately a 66% reduction).

Table 2. Mean marking times in minutes (same teaching staff).

Year	Assignment 1	Assignment 2
2024	16.92	17.89
2025	11.88	6.41

To measure changes in marking accuracy, we compared re-mark requests requiring changes to students' marks from 2024 and 2025 (see Table 3). The number of re-mark requests was roughly the same across both years. While the changes required to marks as a result of these requests dropped for Assignment 2, they rose for Assignment 1. For this assignment, one third of the requests were due to mistakes in manual marking and the other two thirds due to a problem with the autograder taking off marks for the same error twice. Ignoring the latter (the autograder implementation was subsequently modified to fix the error), the percentage drops in 2025 to 11%. All changes to marks for Assignment 2 in 2025 were due to mistakes in manual marking. Aside from the error in the autograder, the fewer mistakes in marking in 2025 was as expected given that there is less manual marking than in 2024.

Table 3. Percentage of re-mark requests resulting in changes to marks.

Year	Assignment 1	Assignment 2
2024	22%	44%
2025	33% (11%)	29%

In terms of students' experience, we compared post-course student evaluations from 2024 and 2025. These anonymous evaluations were carried out separately for undergraduate (CSSE3100) and Masters (CSSE7100) students. Table 4 shows the average score on a 5 point scale (where 5 = Outstanding; 3 = Satisfactory; 1 = Very Poor) for the most relevant questions:

Q4 Course experiences, tools or materials were most useful for my learning.
Q5 Assessment requirements were made clear to me.
Q8 Overall, how would you rate this course?

Table 4. Average student evaluations on scale 5 (= Outstanding) to 1 (= Very Poor).

Year	Q4		Q5		Q8	
	CSSE3100	CSSE7100	CSSE3100	CSSE7100	CSSE3100	CSSE7100
2024	3.95	4.33	4.38	4.33	4.14	4.22
2025	4.49	4.24	4.25	4.54	4.18	4.27

Given the expected variation of such evaluations between cohorts in different years, there does not seem to be any major changes in students' experience of the course related to the autograding. The only significant change is an increase in 2025 for the result of Q4 for undergraduate (CSSE3100) students.

The evaluations also allow the students to add comments. Most of these that referred to assessment or the autograder were positive, including the following:

"Assessments were always marked and returned very quickly which is also much appreciated."

"Lectures, tutorials and assignments all were interesting and valuable learning experiences."

"It was stated that the autograder was new this year: this came as a shock for me. I think it worked quite well and makes sense to have in a course like this (I feel for the markers beforehand)."

7   Conclusion

We have shown how to use Dafny to build an autograder for checking weakest precondition proofs and Dafny specifications. Our autograder supports teaching staff by being general enough to be used for similar assignments (only a few inputs need to be changed from one run of a course to the next), and supports students by providing them with fine-grained feedback (e.g., identifying individual conjuncts of specifications which are missing or wrong) and allowing them maximum flexibility (in terms of programming style) when writing Dafny code.

The autograder was successfully trialed on a cohort of 217 undergraduate and Masters students in 2025. It has been extended for use in a third assignment involving Dafny classes and data abstraction.

Future work could look at using the implemented techniques to provide immediate feedback to students as part of a tutoring system; for example, by providing feedback to short-answer quiz questions intended for learning rather than assessment.

While we would prefer to keep the autogarder code hidden from our current and future students, we are happy to share it with members of the wider teaching community upon request.

References

1. Ahrendt, W., Beckert, B., Bubel, R., Hähnle, R., Schmitt, P.H., Ulbrich, M. (eds.): Deductive Software Verification - The KeY Book - From Theory to Practice. LNCS, vol. 10001. Springer, Cham (2016). https://doi.org/10.1007/978-3-319-49812-6
2. Bertot, Y., Castéran, P.: Interactive Theorem Proving and Program Development - Coq'Art: The Calculus of Inductive Constructions. Texts in Theoretical Computer Science. An EATCS Series. Springer, Heidelberg (2004). https://doi.org/10.1007/978-3-662-07964-5
3. Blazy, S.: Teaching deductive verification in Why3 to undergraduate students. In: Dongol, B., Petre, L., Smith, G. (eds.) FMTea 2019. LNCS, vol. 11758, pp. 52–66. Springer, Cham (2019). https://doi.org/10.1007/978-3-030-32441-4_4
4. Creuse, L., Dross, C., Garion, C., Hugues, J., Huguet, J.: Teaching deductive verification through Frama-C and SPARK for non computer scientists. In: Dongol, B., Petre, L., Smith, G. (eds.) FMTea 2019. LNCS, vol. 11758, pp. 23–36. Springer, Cham (2019). https://doi.org/10.1007/978-3-030-32441-4_2
5. Dijkstra, E.W.: A Discipline of Programming. Prentice-Hall (1976)
6. Dijkstra, E.W., Scholten, C.S.: Predicate Calculus and Program Semantics. Springer, New York (1990). https://doi.org/10.1007/978-1-4612-3228-5
7. Divasón, J., Romero, A.: Using krakatoa for teaching formal verification of Java programs. In: Dongol, B., Petre, L., Smith, G. (eds.) FMTea 2019. LNCS, vol. 11758, pp. 37–51. Springer, Cham (2019). https://doi.org/10.1007/978-3-030-32441-4_3
8. Ettinger, R.: Lessons of formal program design in Dafny. In: Ferreira, J.F., Mendes, A., Menghi, C. (eds.) FMTea 2021. LNCS, vol. 13122, pp. 84–100. Springer, Cham (2021). https://doi.org/10.1007/978-3-030-91550-6_7
9. Filliâtre, J.-C., Paskevich, A.: Why3 — where programs meet provers. In: Felleisen, M., Gardner, P. (eds.) ESOP 2013. LNCS, vol. 7792, pp. 125–128. Springer, Heidelberg (2013). https://doi.org/10.1007/978-3-642-37036-6_8
10. Hamza, J., Voirol, N., Kunčak, V.: System FR: formalized foundations for the Stainless verifier. Proc. ACM Program. Lang. 3(OOPSLA), 166:1–166:30 (2019). https://doi.org/10.1145/3360592
11. Hoare, C.A.R.: An axiomatic basis for computer programming. Commun. ACM 12(10), 576–580 (1969). https://doi.org/10.1145/363235.363259
12. Leino, K.R.M.: Dafny: an automatic program verifier for functional correctness. In: Clarke, E.M., Voronkov, A. (eds.) LPAR 2010. LNCS (LNAI), vol. 6355, pp. 348–370. Springer, Heidelberg (2010). https://doi.org/10.1007/978-3-642-17511-4_20
13. Leino, K.R.M.: Program Proofs. MIT Press (2023)

14. Leino, K.R.M., Pit-Claudel, C.: Trigger selection strategies to stabilize program verifiers. In: Chaudhuri, S., Farzan, A. (eds.) CAV 2016. LNCS, vol. 9779, pp. 361–381. Springer, Cham (2016). https://doi.org/10.1007/978-3-319-41528-4_20
15. Leino, K.R.M., Polikarpova, N.: Verified calculations. In: Cohen, E., Rybalchenko, A. (eds.) VSTTE 2013. LNCS, vol. 8164, pp. 170–190. Springer, Heidelberg (2014). https://doi.org/10.1007/978-3-642-54108-7_9
16. Milovančević, D., Bucev, M., Wojnarowski, M., Chassot, S., Kunčak, V.: Formal autograding in a classroom. In: Vafeiadis, V. (ed.) Programming Languages and Systems - 34th European Symposium on Programming, ESOP 2025. LNCS, vol. 15695, pp. 154–174. Springer, Heidelberg (2025). https://doi.org/10.1007/978-3-031-91121-7_7
17. Noble, J., Streader, D., Gariano, I.O., Samarakoon, M.: More programming than programming: teaching formal methods in a software engineering programme. In: Deshmukh, J.V., Havelund, K., Perez, I. (eds.) 14th International Symposium on NASA Formal Methods, NFM 2022. LNCS, vol. 13260, pp. 431–450. Springer, Heidelberg (2022). https://doi.org/10.1007/978-3-031-06773-0_23
18. Pierce, B.C., et al.: Software foundations, **16** (2010). http://softwarefoundations.cis.upenn.edu/

Automatic Assessment and Feedback on Undergraduates' Structural Induction Proofs

Edward Sabinus[(✉)][ID], Thomas Kühn[ID], and Wolf Zimmermann

Martin-Luther-University Halle-Wittenberg, 06120 Halle, Germany
{edward.sabinus,thomas.kuehn,wolf.zimmermann}@informatik.uni-halle.de

Abstract. To help students learn the structural induction proof technique, we aim to provide a scalable approach that delivers immediate, comprehensive feedback as well as a fair assessment for students' proofs. To this end, we developed an automatic proof checking approach tailored for helpful feedback, a fair assessment schema for structural induction proofs, and a simple DSL for tasks and solutions. We implemented Automatic Structural Induction Assessor (ASIA), a command line tool that automatically provides helpful feedback and awards points for a given solution. To evaluate our approach in practice, we employed ASIA in three undergraduate computer science courses (with around 60 students each) and conducted a questionnaire survey focusing on the tool's usability, quality of the provided feedback and the student's work load. We found that ASIA provided comprehensive feedback and a suitable assessment for students, as well as limited manual reassessment for teachers.

Keywords: Automatic Assessment · Structural Induction · Teaching Feedback · Abstract Data Types

1 Introduction

In undergraduate computer science education, proof techniques are difficult to learn, yet they are a foundational topic. In particular, structural induction is difficult to convey due to the gap between knowing the principle and applying it in practice [23]. However, to effectively train the application of structural induction in practice, students must train it a lot and get feedback quickly. While providing additional examples is easy, delivering comprehensible feedback does not scale well with the number of students in undergraduate computer science courses. On top of that, assessing structural induction proofs is more difficult because there might be a wide variety of correct solutions the assessor must manually evaluate. Consequently, when many different assessors are employed, they tend to assess the students' solutions unfairly. To increase fairness without reducing feedback quality, a single automated assessor should evaluate all solutions. Nevertheless, to the best of our knowledge, there are currently no automated methods for assessing structural induction proofs (cf. Sect. 2). Although existing proof checkers like Isabelle/HOL [18], Rocq [3] (formerly known as Coq), and PVS [19] can determine whether a proof is correct or not, but they can neither give detailed feedback on mistakes nor grade the proof, thus, their feedback would not be helpful for students. There is the LogInd tool [14], which can provide hints and feedback to students' structural induction proofs, but cannot grade those proofs. Then there is CalcCheck [12], which provides feedback and assessment, but is limited to mathematical induction proofs, while structural induction applies to arbitrary data types.

Our main goal is to provide immediate feedback and automatic assessment of structural induction proofs. For this, we answer the following research questions:

RQ1 How must proof checking be performed to generate comprehensible feedback for students?

RQ2 How can erroneous solutions be automatically assessed, such that the assessment is correct and fair, as well as graded according to errors?

RQ3 How suitable is automatic assessment in an undergraduate course setting?

RQ4 How usable is the resulting tool for undergraduate students?

To answer these questions, we designed a representation (Sect. 4) for structural induction tasks, their solutions, and their underlying foundations (Sect. 3), which in our course are *abstract data types* (ADT). Moreover, to provide helpful feedback (RQ1), proof checking must consider follow-up errors. To this end, we first identify possible errors and assign them to the minimal context of the proof, which is minimal for recognizing that error. The proof checking examines these errors separately for each context and marks them in the model. Then, the assessment (Sect. 5) evaluates each error separately and aggregates the final assessment (RQ2). Our main contributions are:

- Immediate comprehensible feedback for structural induction proofs,
- Fair and gradual assessment schema for structural induction proofs, and
- A practical and feasible tool for teaching structural induction to students.

We integrated ASIA into the local online exercise platform YAPEX [7] for our course. We evaluated (Sect. 7) the general suitability of ASIA (RQ3) by employing it in the case study of three consecutive undergraduate courses (2023–2025) on computer science with around 60 active students each year. Moreover, to study the usability of ASIA for students (RQ4), we conducted a questionnaire survey in 2025, receiving 27 responses. The survey was split into three parts focusing on the general usability of ASIA in YAPEX, employing the *System Usability Scale* (SUS) [4], the perceived feedback quality, as well as the perceived work load assessed via the *NASA Task Load Index* (TLX) [8]. The SUS is a standardized questionnaire to assess the subjective usability of a system. Likewise, TLX is a standardized, subjective, multidimensional assessment scheme rating the perceived workload when performing a given task. Both result in a normalized score (ranging from 0 to 100) of the perceived usability of a system respectively work load when using the system. We found that ASIA could automatically assess most of the students' assignments, whereas only a limited

Table 1. Comparison of related approaches for assessing formal proofs

Criterion	Rocq [3]	Isabelle/HOL [18]	PVS [19]	LogInd [14]	CalcCheck [12]	Ωmega [1]	eMathChecker [2]	ComIn-M [21]	Yalep [15]	LeanTutor [20]	SIETTE [6]	AI autograders [25]
Proof checking	■	■	■	■	■	■	■	■	■	■	□	■
Mathematical induction	■	■	■	■	■	■	■	■	■	■	□	■
Structural induction	■	■	■	■	■	□	□	□	□	□	□	□
Any abstract data types	■	■	■	□	□	□	□	□	□	□	□	□
Assignment checking	□	□	□	■	■	■	■	■	■	■	■	■
Follow-up error consideration	□	□	□	■	■	■	■	■	■	■	□	■
Automatic assessment	□	□	□	□	■	□	□	□	□	□	■	■
Reliable Correctness	■	■	■	■	■	■	■	■	■	■	■	□

■ supported   □ not-supported

number of submissions had to be manually reassessed. Moreover, our evaluation suggests that most participants perceived the feedback as precise, helpful, and easy to understand. Furthermore, we found that ASIA's integration into YAPEX indicates an above average usability ($SUS\ score = 62.5$), as well as a reduced work load ($raw\ TLX = 40$) for students. The benefits are a scalable method for teaching structural induction proofs to undergraduate students providing immediate, comprehensible feedback to students and a fair, automatic assessment for teachers. In [22], we provide the replication package, containing ASIA, featured exercises, as well as anonymized questionnaire and case study results.

2   State of the Art in Assessing Formal Proofs

Henceforth, we provide an overview on existing tools for assessing formal proofs, whereas the comparison is illustrated in Table 1. We considered approaches ranging from proof checkers to mathematical tutoring systems. While most approaches are able to check proofs that employ mathematical induction, only [3,12,14,18,19] support structural induction proofs. From those, only [3,18,19] allow teachers to specify arbitrary abstract data types. In contrast to them, most tools able to assess student assignments take follow-up errors into account, i.e., continue the assessment, although a proof step is incorrect. Notably, only [6,12,25] perform a fully automatic assessment (i.e., assess and mark an assignment). However, as LLM-based approaches can be incorrect when assessing a student's proof, in contrast to formal approaches, we consider the correctness of [25] as unreliable.

In detail, proof checkers, like Isabelle/HOL [18], Rocq [3] (formerly known as Coq), and PVS [19], have a strong logical foundation and thus can check any kind of proof. However, they are less effective in giving feedback to students. More precisely, they can only determine if a given proof is correct or not. If not, they provide the resulting state of the proof, showing what should be proven further to prove the goal. Mistakes in the proof are either ignored and the proof marked as incomplete (Rocq), or the whole proof, beginning at the first mistake, gets marked as wrong (Isabelle/HOL). Consequently, proof checkers do not consider follow-up errors. Thus, students would barely be able to suggest a fix for the mistake. Such proof checking is neither useful for the generation of helpful feedback for students nor for an assessment.

However, tutoring systems for structural induction exist, like the LogInd tool [14]. LogInd is designed to give students hints and feedback on their structural induction proofs. Students must click on a *submit* button after each step to continue the proof and get feedback. Thus, this method is only applicable for practicing, but not for exams or assessments. LogInd uses a domain reasoner in the background to give students the freedom to design their own proofs and give them hints and feedback. Currently, implemented domains are the structure of propositional logic formulas and natural numbers [14]. Thus, it cannot assist students on arbitrary structures (abstract data types), and any new structure would need a new domain reasoner, making it hard to create new exercises with new structures. Another tool designed to provide students with feedback on their proofs is CalcCheck [12], which can check mathematical proofs, including induction over natural numbers and sequences. It is designed to give feedback to students and can also calculate an assessment for those proofs. However, it cannot check or assess structural induction proofs over arbitrary inductive data types. For mathematical induction, several works [1,2,15,20,21] focus on educational proof checking and feedback. However, none of them can calculate an assessment by means of a number of points, nor check induction over arbitrary inductive data types. Most of them employ a proof checker in the background, e.g., [2,15,20], while *LeanTutor* [20] uses an LLM to communicate between students and a proof checker. While the above approaches ensure a correct proof assessment, proof assessment tools using artificial intelligence, such as, *AI autograders* [25], rely on LLMs to check and assess the proof, which can be unreliable due to hallucinations and imprecisions. In [25], the authors report a grading accuracy of about $80-90\%$, which is more accurate than humans, yet still cannot match the 100% accuracy of formal assessment tools.

In general, automatic assessments are more common for other types of exercises. The most common automatically assessed tasks are single-choice, multiple-choice, and cloze questions, like in [6]. Moreover, programming exercises are often automatically assessed [9]. While most of them can test submitted programs, e.g., using black box and/or unit tests, they cannot assess them (i.e., award points). In contrast, *JavAssess* [10] tests and assesses (marks) student programs. Similarly, *Dr. Scratch* [17] uses program analysis to verify and assess Scratch programs. We refer the interested reader to [9] for a more thorough comparison. We aim to create a tool able to reliably check proofs over arbitrary abstract data types, supporting both mathematical and structural induction. Moreover, it should be able to automatically assess proof assignments, taking follow-up errors into account.

3   Structural Induction Proofs on Abstract Data Types

As the basis for our structural induction proofs, we use abstract data types (ADT) [24]. To make structural induction proofs we also need direct proofs, which are required to prove each induction case.

An ADT [24] has sorts, operations, variables, and axioms. In Example 1, the ADT has exactly one sort ($\mathbb{N}$), in general there can be more than one sort. The operations are $0, inc, +$, where $0, inc$ are constructors required for structural induction proofs. In general, sorts can have an arbitrary number of constructors. Axioms, such as A1 and A2, are equations over the operations and variables.

Example 1. Abstract Data Type of the Peano arithmetic
Natural numbers $\mathbb{N}$ are inductively defined as: $0 \in \mathbb{N}$ and $\forall n \in \mathbb{N} : inc(n) \in \mathbb{N}$.
The addition $+ : \mathbb{N} \times \mathbb{N} \to \mathbb{N}$ is defined by the following axioms
A1: $\forall n \in \mathbb{N} : \qquad\qquad n + 0 = n$

A2: $\forall n, m \in \mathbb{N} : \quad n + inc(m) = inc(n + m)$

A direct proof is a sequence of term rewritings. Each term rewriting applies a rule, e.g. axiom. To each term rewriting belongs the applied rule, the application's direction, the subterm, on which the rule is applied, and the substitution of the rule, to match the subterm.

Example 2. Task and direct proof in the Peano arithmetic
Task: $\forall n \in \mathbb{N} : n + inc(0) = inc(n)$
Proof: $\underline{n + inc(0)} =^{A2,[0/m]} inc(\underline{n + 0}) =^{A1,[]} inc(n)$

Example 2 shows a task and a direct proof. The task consists of the equation and the variables used. The proof has two term rewritings: It starts with the left side of the task's equation. Axiom A2 is applied with substitution $[0/m]$ on the subterm $n + inc(0)$. Then axiom A1 is applied without substitution on subterm $n + 0$. The result is the right side of the task's equation.

With the structural induction principle [5] the proof iterates over the constructors of the induction variable. In each iteration (in this paper called the induction case), the induction variable gets replaced by a minimal term with the corresponding constructor. If the constructor has no building parameter, the induction case is called induction basis (IB), else induction step (IS). Once formulated, the induction hypothesis (IH) is applicable to induction steps. When using the induction hypothesis, there are fixed variables that cannot be substituted. When a constructor has multiple building parameters, multiple induction hypotheses can be required.

Example 3. Task and proof with structural induction.
Task: $\forall n \in \mathbb{N} : 0 + n = n$, induction n
IB: $0 + 0 = 0$ Proof: $\underline{0 + 0} =^{A1,[0/n]} 0$
IH: for one fixed $n \in \mathbb{N} : 0 + n = n$
IS: for one fixed $n \in \mathbb{N} : 0 + n = n \rightarrow 0 + inc(n) = inc(n)$
Proof: $\underline{0 + inc(n)} =^{A2,[0/n,n/m]} inc(\underline{0 + n}) =^{IH,[]} inc(n)$

Example 3 shows a task and a structural induction proof. The task contains additionally an induction variable. The proof consists of three parts: IBs, IHs, and ISs. Each induction case starts with its goal, followed by a direct proof. Each induction hypothesis consists of an equation with its variables. First, the induction variable n is replaced with its basic constructor 0 in IB. Then, as we have proven all IBs, we can formulate the IH, which assumes the goal for one fixed n. With IS we replace n with its building constructor $inc(n)$ and show this goal, while assuming IH.

4   Proof Checking and Representation

To enable the proof checking, we first need to represent the proof and additionally the abstract data type and the task, as described in Sect. 3. The grammars in Fig. 1 specify the domain-specific languages for abstract data types (ADT), tasks, and proofs as grammars in EBNF [11].

The abstract data type (Fig. 1a) consists of a list of sorts, their constructing operations, other operation signatures, variable definitions, and axioms. The task (Fig. 1b) consists of an equation, a configuration for the assessment, and optionally the induction variable. For an induction task, the configuration consists of multiple subtasks for each constructor of the induction variable. The proof (Fig. 1c) can be either an induction proof or a direct proof. The direct proof starts with one term and applies a sequence of term rewritings, each of which consists of a rule that is applied (e.g. an axiom), the direction of the application (i.e. from left to right or right to left), the sub term on which the rule is applied, the substitution required for the application, and the resulting term. The induction proof can define the induction variable and consists of one or more base cases and induction hypotheses, and can have a list of induction steps. Each base case and induction step is a direct proof.

As discussed in Sect. 2, follow-up error consideration is essential for proof checking tailored to feedback generation (RQ1). The key idea to achieve proof checking with follow-up error consideration is to enumerate all possible errors, assign them to a minimal proof's context, check each error separately, and mark all found errors in the proof. Due to the high number of concrete error types, we will not enumerate all and abstract the concepts from them. All errors are separated into three error kinds: formal errors, content errors, and warnings. Formal errors violate the definition of the context and make further assessment of that context pointless and impossible. Content errors make the proof invalid, while they do not violate the formalities of the proof. Warnings are regarding unnecessary information in the proof that does not invalidate the proof, but can confuse human assessors.

(a) Grammar of abstract data types

```
ADT          ::= Sorts Constructors? Operations Vars Axioms?
Sorts        ::= 'sorts' Sort (',' Sort)*
Constructors ::= 'constructors' Operation+
Operations   ::= 'operations' Operation+
Operation    ::= OperationName ':' (Sort ('><' Sort)* '->')? Sort
Vars         ::= Var ':' Sort (',' Var ':' Sort)*
Axioms       ::= 'axioms' (AxiomName ':' Term '=' Term)+
Term         ::= Var | (OperationName ('(' Term (',' Term)* ')')?)
```

(b) Grammar of a task

```
Task      ::= 'task' Eq ('induction' Var)? (TaskPt | SubTasks)
Eq        ::= (Fixed ':')? (Forall ':')? Term '=' Term
Fixed     ::= 'fixed' Var ':' Sort (',' Var ':' Sort)*
Forall    ::= 'forall' Var ':' Sort (',' Var ':' Sort)*
TaskPt    ::= 'maxpt' Nat 'minsteps' Nat 'maxsteps' Nat
SubTasks  ::= ('case' ConstructorName TaskPt)+ 'IH' 'maxpt' Nat
```

(c) Grammar of direct and induction proofs

```
Proof           ::= 'proof' DirectProof | InductionProof
InductionProof  ::= ('induction' Var)? Induction
Induction       ::= Basis+ Hypothesis+ Step*
Basis           ::= 'IB' Goal DirectProof
Hypothesis      ::= 'IH' IHName? Eq
Step            ::= 'IS' Goal DirectProof
Goal            ::= 'goal :' Eq
DirectProof     ::= Term TermRewriting+
TermRewriting   ::= '{' Rule ',' ('lr' | 'rl') ',' Term ',' Subst '}' '=' Term
Subst           ::= '[' ( Term '/' Var (',' Term '/' Var)*)? ']'
```

Fig. 1. EBNF [11] grammars for abstract data types (a), tasks (b), and proofs (c). Non-terminal symbols that are not defined, like Sort, OperationName, or Var are identifiers, except for Nat, which is a natural number.

Listing 1 shows an erroneous proof, we will use to illustrate the errors in it. In this proof, the induction basis (Line 2–5) and induction step (Line 7–9) are mixed. The first term rewriting in Line 4 is correct, yet in the second rewriting in Line 5, the axiom matches only the right side of the equation, i.e. the given direction and substitution are wrong. In the term rewriting in Line 9, the application of the induction hypothesis IH in the induction basis and the substitution of the induction variable n is forbidden.

In sum, to construct comprehensible feedback on the basis of proof checking (RQ1), we consider follow-up errors in erroneous solutions by checking every error within its smallest context in the proof, independently from previous errors. This enables the detection of additional errors and the verification of later parts of the proof as correct. This follow-up error consideration enables comprehensible feedback in contrast to proof checkers (c.f. Sect. 2).

Listing 1. Student solution for the proof from Example 3.

```
 1 proof
 2 IB: goal: fixed n:Nat : add(zero, inc(n)) = inc(n)
 3 add(zero, inc(n))
 4   {A2, lr, add(zero,inc(n)), [zero/n, n/m]}
 5 = inc(add(zero, n))
 6   {A1, lr, add(zero,n)       , []              }
 7 = inc(n)
 8 IH: fixed n:Nat : add(zero, n) = n
 9 IS: goal: add(zero, zero) = zero
10 add(zero, zero)
11   {IH, lr, add(zero, zero) , [zero/n]       }
12 = zero
```

5   Assessment Schema

To provide an assessment scheme that is correct and fair, as well as graded according to errors (RQ2), we propose a deductive grading scheme whereas the assessment deducts points for mistakes the student makes.

Let $\hat{F}$ and $\hat{C}$ be the disjoint sets of formal and content errors, as well as $\hat{E} = \hat{F} \uplus \hat{C}$ the set of all errors. A proof is either a direct proof dp or a structural induction ind, as defined in Fig. 2. A direct proof $dp = (pt, l, u, \langle tr_1, \ldots, tr_n \rangle)$ is a tupel, with the maximal number of points pt, the lower bound (minimal number) of steps l and upper bound (maximal number) of steps u,

$$
\varphi(p) = \begin{cases}
0 & \textbf{if } (p \equiv tr_i \vee p \equiv ih) \\ & \wedge F_p \neq \emptyset & (1) \\[2mm]
1 - \sum_{c \in C_p} \alpha(c) & \textbf{if } (p \equiv tr_i \vee p \equiv ih) \\ & \wedge F_p = \emptyset & (2) \\[2mm]
pt \cdot \min\left(1, \frac{n}{l}\right) \cdot \left(\sum_{i=1}^{n} \frac{\varphi(tr_i)}{\min(n, u)}\right) & \textbf{if } p \equiv dp & (3) \\[2mm]
0 & \textbf{if } p \equiv case \\ & \wedge F_{goal} \neq \emptyset & (4) \\[2mm]
pt \cdot \min\left(1, \frac{n}{l}\right) \cdot \frac{\sum_{i=1}^{n} \left(\varphi(tr_i) - \sum_{c \in C_{goal}} \alpha(c)\right)}{\min(n, u)} & \textbf{if } p \equiv case \\ & \wedge F_{goal} = \emptyset & (5) \\[2mm]
pt_{ih} \cdot \varphi(ih) + \sum_{i=1}^{m} \varphi(case_i) & \textbf{if } p \equiv ind & (6)
\end{cases}
$$

Fig. 2. Formal definition of the assessment function $\varphi : P \to \mathbb{Q}_0^+$. Here, a direct proof $dp = (pt, l, u, \langle tr_1, \cdots, tr_n \rangle)$ consists of the maximal points pt, the lower l and upper u bound of steps, and the term rewritings tr_i (with $i \in \{1, \ldots, n\}$); a $case = (goal, pt, l, u, \langle tr_1, \cdots, tr_n \rangle)$ additionally includes the $goal$, whereas an induction proof $ind = (pt_{ih}, ih, \{case_1, \ldots, case_m\})$ encompasses all cases, the induction hypothesis ih and corresponding points pt_{ih}.

and a sequence of term rewritings tr_i (with $i \in \{1, \ldots, n\}$). An induction $ind = (pt_{ih}, ih, \{case_1, \ldots, case_m\})$ has the points awarded for the induction hypothesis as first component, followed by the induction hypothesis ih and the set of induction cases $\{case_1, \ldots, case_m\}$. In contrast to a direct proof, each $case = (goal, pt, l, u, \langle tr_1, \ldots, tr_n \rangle)$ is additionally characterized by a proof $goal$, i.e., the equation to be shown. Let P be the set of all direct proofs, structural inductions, and their parts, i.e., term rewritings, direct proofs, inductions, induction hypotheses, cases, proof goals.

As the possible errors depend on the part p of the proof under investigation, $\hat{E}_p \subseteq \hat{E}$ denotes the set of possible errors in part p, e.g., $\hat{E}_{tr}$ denotes the set of possible errors in a term rewriting. Likewise, $\hat{F}_p \subseteq \hat{F}$ and $\hat{C}_p \subseteq \hat{C}$ denotes the set of possible formal respectively content errors in part p of the proof. Similarly, $E_p \subseteq \hat{E}_p$ denotes the set of actually found errors found in part p, e.g. $E_{tr} = \{c_4, c_5\}$ denotes that the term rewriting tr contains the content errors c_4 and c_5. Conversely, $F_p \subseteq \hat{F}_p$ and $C_p \subseteq \hat{C}_p$ denotes the set of actual formal respectively content errors found in part p of the proof. The weighting function $\alpha : C \to \mathbb{Q}_0^+$ assigns a weight $q \in \mathbb{Q}_0^+$ with $0 \leq q \leq 1$ to each content error, to configure the penalty each content error incourse. To guarantee, that the assessment $\varphi(p) \geq 0$ for each part of a proof p, we assume that for each p it holds that $\sum_{c \in C_p} \alpha(c) = 1$. Finally, we can define the recursive assessment function $\varphi : P \to \mathbb{Q}_0^+$ as depicted in Fig. 2.

Without loss of generality, we assume that if the result of the assessment is negative, then it is truncated to zero. Equations 1–3 show the assessment of a direct proof dp. The assessment's maximal points pt are down scaled if the minimal steps l are not reached. Furthermore, the assessment is scaled if the maximal steps u are exceeded, such that the error points should not be infinitely down scalable. Besides that, the assessment of a direct proof is the sum of the assessment of its term replacements. If formal errors F occur, the assessment is zero. Otherwise, for term rewritings, the sum of the configured weights α is subtracted from one abstract assessment point. For grading structural inductions, the assessment function is extended by Equations 4–6. The assessment of the structural induction ind is the sum of the assessments of the induction hypothesis and the induction cases. Induction cases are induction base cases and induction steps. The induction hypothesis has its own maximal points pt and is assessed in the same way as a term rewriting. Each induction case is assessed as direct proof with an additionally assessed proof goal.

Example 4. Assessment of the induction case `add(zero,inc(n))=inc(n)`
Let $inc = (inc_{goal}, 2, 2, 2, \langle tr_1, tr_2 \rangle)$ be the first induction case from Listing 1, whereas $|C_{goal}| = 1$, $C_{tr_1} = \emptyset$ and $|C_{tr_2}| = 2$. For simplicity, we assume equal

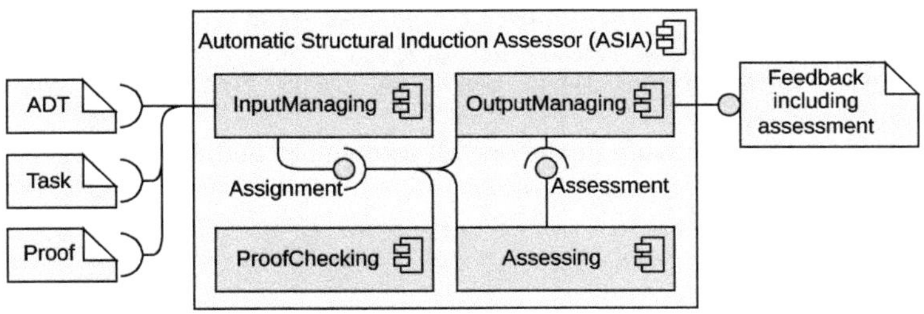

Fig. 3. Architecture of ASIA as UML component diagram, where ADT, Task and Proof are input files and the feedback including assessment is its output.

error weights $\alpha(c) = 0.25$ for all content errors c.

$$\varphi(inc) = 2 \cdot \min\left(1, \frac{2}{2}\right) \cdot \frac{\sum_{i=1}^{2}\left(\varphi(tr_i) - \sum_{c \in C_{goal}} \alpha(c)\right)}{\min(2,2)}$$

$$= \sum_{i=1}^{2}\left(\varphi(tr_i) - \sum_{c \in C_{goal}} \alpha(c)\right) = (\varphi(tr_1) - 0.25) + (\varphi(tr_2) - 0.25)$$

$$= \left(1 - \left(\sum_{c \in C_{tr_1}} \alpha(c)\right) - 0.25\right) + \left(1 - \left(\sum_{c \in C_{tr_2}} \alpha(c)\right) - 0.25\right)$$

$$= (1 - 0 - 0.25) + (1 - (0.25 + 0.25) - 0.25) = 0.75 + 0.25 = \underline{1.0}$$

In summary, to assess erroneous solutions automatically, such that the assessment is correct and fair, as well as graded according to errors (RQ2), we created the assessment schema which calculates the assessment based on the found errors. The approach is fair by construction, as every student's solution is objectively assessed using the same formula, which is independent from the student or the assessors mood, as it would be with human assessors.

6   Automatic Structural Induction Assessor (ASIA)

Here, we give a brief overview of ASIA's implementation (for details, cf. [22]) describing its architecture, discussing the kinds of feedback provided, and highlighting the employed online exercise platform YAPEX.

Architecture. ASIA is implemented as a command-line tool written in $C\#$, which reads the ADT, task, and proof from given files. Its architecture is divided into four components, as shown in Fig. 3. InputManaging parses the files and constructs an internal representation. The ProofChecking component checks the proof marking mistakes in this representation, whereas the Assessing component calculates the assessment (cf. Sect. 5) based on the mistakes. Finally, OutputManaging prints the assessment with feedback onto the command-line.

(a) Reporting of syntax error

```
1 | Error in 4:3: The rule Blub is not defined
2 | Error in 4:6: mismatched input 'leftright' expecting {'lr', 'rl'}
```

(b) Reporting of the overall assessment and feedback

```
1 | Overall Assessment: 2 of 4
2 |   Case zero: 0 of 1
3 |   IH: 1 of 1
4 |   Case inc: 1 of 2
5 | Errors in (2:0) case inc:
6 |    41 Kind of induction case inconsistent to constructor
7 | Errors in (6:2) case inc, term rewriting 2:
8 |     8 Wrong direction
9 |    10 Wrong substitution at wrong direction. Hint:[ add( zero,n ) / n ]
```

Fig. 4. Example feedback provided by ASIA in case of syntax errors (a) and the given assessment and feedback (b) for the proof in Listing 1.

Feedback. Our tool gives three kinds of feedback: syntax errors and proof-checking errors, and the assessment. *Syntax errors* are reported, if the input does not conform to the grammar or contains undefined terms of the ADT. In case of syntax errors, ASIA aborts the proof checking and assessment. Figure 4a shows two syntax errors. Line 1 reports the use of an undefined axiom or induction hypothesis, whereas Line 2 reports that the rewriting direction must be either rl or lr. In each case, ASIA provides the precise error location and corresponding error message. In contrast, *proof-checking errors* are detected during ProofChecking. These additionally provide their position within the proof structure, e.g., which term rewriting, to help students understand the context of the error, as illustrated in Fig. 4. Furthermore, for each error, ASIA prints an error code and a short description of the error. In some cases ASIA provides hints to students, e.g., hinting a correct substitution in Line 9. All of this information provided to students is working towards the comprehensibility of the mistakes in the proof. While all error messages provide feedback to students, the *assessment* gives an overview of the points awarded by ASIA. As highlighted in Lines 1–4 of Listing 1, this overview not only comprises the overall awarded and achievable points but also encompasses the assessment of each subtask.

YAPEX. Our students already use the online exercise platform YAPEX [7] for courses, like programming. It has a user-friendly interface with a built-in editor and ability to run tests on the students' input, e.g., programs. As ASIA has a simple command-line interface, it was easy to integrate into YAPEX, such that students could write their proofs in the editor, check those proofs with ASIA by a button press, and get the assessment and feedback from ASIA displayed.

7   Evaluation of ASIA in an Undergraduate Course

Methodology. For evaluating ASIA's suitability (RQ3), we collected data on its use in an undergraduate course on *mathematical foundations of computer* *science and modeling concepts* at MLU in three consecutive years: 2023, 2024 and 2025. Each year, students received five separate assignments via YAPEX, to prove a statement over an ADT using structural induction with ASIA, where they could only check for syntax errors. Before each assignment, we provide training exercises, that allow for students to receive immediate feedback on their proofs, including *syntax* and *proof-checking errors*. To study its suitability for the automatic assessment of students' submissions, we collected both the automatic assessment and the manual reassessment performed by three student research assistants each year. Their task was to reassess all submissions where points were deduced. For each submission, they fixed small syntactical errors, e.g., missing/misplaced brackets, commented out incomplete induction cases, until ASIA could assess it. In case the submission did not adhere to the proof's syntax, an assistant manually graded the submission using the same grading scheme.

To evaluate ASIA's usability and effects on students learning a new proof technique (RQ4), we performed a questionnaire survey among the students in 2025. We invited all corresponding course students for voluntary, anonymous participation. The questionnaire encompassed three parts. First, we evaluated the perceived usability of ASIA by employing the *System Usability Scale* [4] encompassing ten subjective statements on the system's usage context, whereas students' agreement to each was rated on a 5-point Likert scale. Second, we surveyed the perceived quality of the feedback provided by ASIA. To this end, we derived six statements about the feedback quality from the guidelines for good error messages [16]. The agreement to each statement was again rated on a 5-point Likert scale. Third, we evaluated the tool's impact on the students' work load compared to a handwritten proof by employing the *NASA Task Load Index* [8]. Here, students rate their perceived load in six categories, e.g., mental demand or own effort, using a 21-point scale. We refer the interested reader to [8] for more details. The questionnaire is available in our replication package [22].

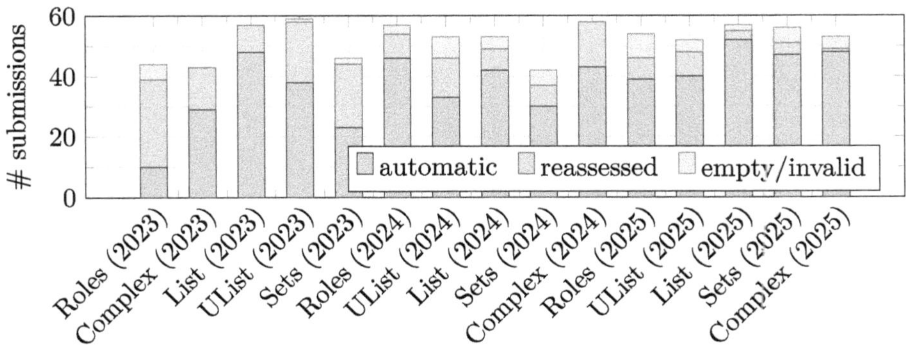

Fig. 5. Case study spanning annual courses from 2023 to 2025. The assignments on structural inductions are shown in order of their assignment, highlighting the reachable points in brackets. The bars indicate the amount of automatically assessed solutions, manually reassessed solutions, and empty or invalid solutions.

Participants. The course is taught in the second semester and had 68 participants in 2023, 65 in 2024, and 59 in 2025. About half of the semester students use ASIA for exercises. As students were not required to complete all assignments, the number of submissions shown in Fig. 5 varies for each of the five assignments. While the order of the assignments changed from 2023 to 2024, the assignments remained unchanged. Similarly, 2 of the 3 student research assistants reassessed in all 3 years employing the same assessment schema. For our suitability evaluation, we consider only valid submission, i.e., ignoring empty and invalid submissions. Regarding our questionnaire conducted in 2025, 27 (46%) of the 59 course participants responded, whereas 17 (29%) completed the questionnaire.

Results. Regarding the evaluation of ASIA's suitability (RQ3), Fig. 5 shows the total number of automatically assessed solutions, i.e., those receiving maximal points or remained unchanged by the manual reassessment. In 2023, especially in the first assignment, 29 tasks required reassessment. In later exercises, most students were able to submit syntactically correct proofs. In 2024 and 2025, we increased our efforts teaching the DSL to our students, e.g., by providing a wiki outlining the tool's syntax and additional training exercises providing immediate feedback. This gradually reduced the number of reassessments required to one in the last assignment in 2025. Focusing on manually reassessed submissions, Fig. 6 highlights the distribution of the normalized assessment before and after reassessment for each assignment. Moreover, it shows how these assessments differ, indicating that most reassessed solutions in 2023 were only awarded below 25% of the reachable points, whereas the manual reassessment increased the points by a significant amount. In contrast, when considering the years 2024 and 2025, the median automatic assessment is only slightly higher than the median of the manual reassessment. Notably, the amount of reassessment required was reduced over the three years. In turn, Fig. 7 shows the distribution of all auto-

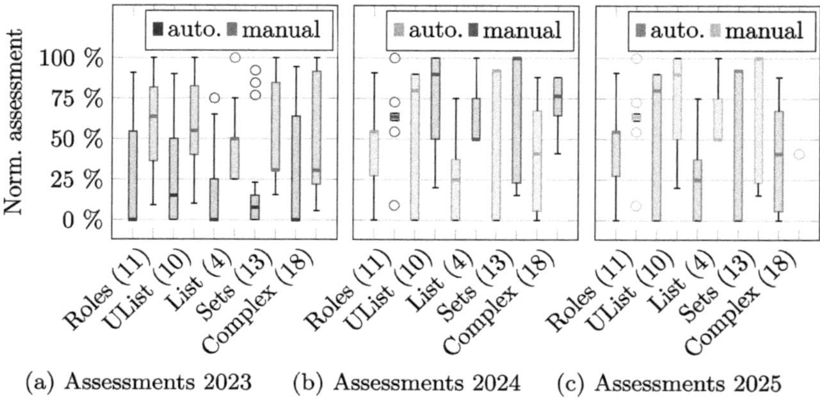

(a) Assessments 2023   (b) Assessments 2024   (c) Assessments 2025

Fig. 6. Box plots of all reassessed submissions for the five assignments in 2023 (a), 2024 (b), and 2025 (c) showing the normalized automatic assessment and manual reassessment, whereas the reachable points are shown in brackets.

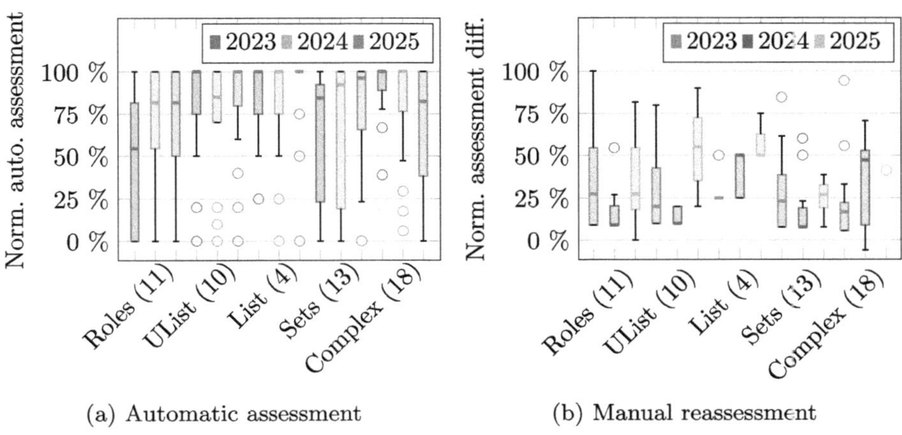

(a) Automatic assessment   (b) Manual reassessment

Fig. 7. Box plots of automatically assessed submissions (a) showing the normalized distribution of assessments and all manually reassessed submissions (b) with the normalized difference between automatic and manual reassessment.

matically assessed solutions. Here, we observe that students adhering to our syntax receive more than 75% of points, except for the first assignment *Roles* in 2023. Apparently, ASIA was able to grade students' submissions gradually.

Aggregating the results from Figs. 6 and 7 depicts the distribution of normalized improvements through manual reassessment. The median improvement through reassessment was limited to 25%, except for the second assignment *UList* in 2025 and the last assignment *Complex* in 2024, where the median improvement was 50%. Please note that we found many outliers when investigating the normalized improvement of reassessments. In particular, for one submission of *Complex* in 2024, the reassessment awarded one point less than ASIA.

In regard to ASIA's usability (RQ4), Fig. 8 summarizes the results of three parts of our questionnaire. First, Fig. 8a depicts the distribution of the agreement with the ten statements of the SUS. The scale alternates positive (teal) and negative statements (orange). Most participants agreed with positive statements, while disagreeing with negative ones. Yet, students' agreement was mixed for the statements *cumbersome to use* and *required a lot of learning to use*. Second, Fig. 8b plots the agreement of the participants with our six statements on the feedback quality (RQ1). Half of the participants agreed with all of them, indicating that the feedback was concise, helpful and comprehensive. Still, some participants did not agree that the feedback explained the cause of the mistake made. Third, Fig. 8c depicts the participants' answers to the six statements of the TLX (RQ4). Students perceived performing structural induction proofs more mentally demanding than physically or time-consuming. Note that the scale of *own performance* in TLX is reversed. Thus, the box plot indicates that most students perceived their own performance as good to very good. Still most students characterized their *own effort* and *frustration* as medium to high. In sum,

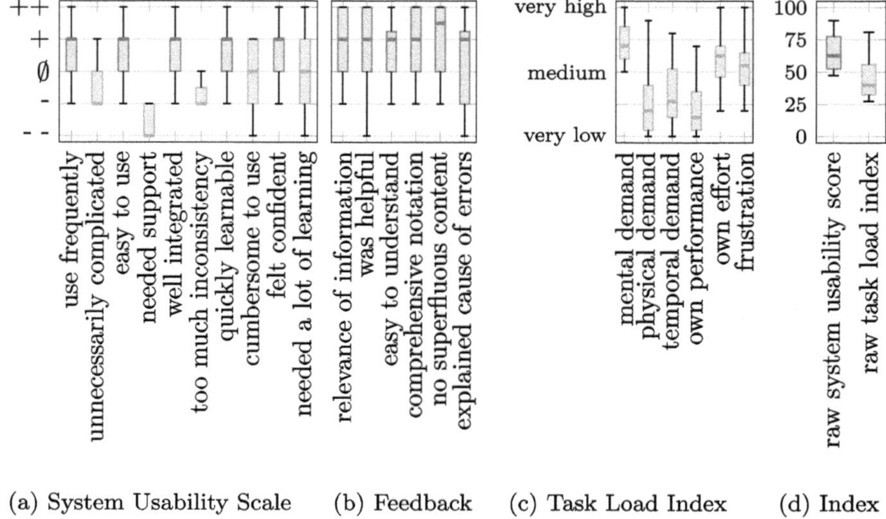

(a) System Usability Scale   (b) Feedback   (c) Task Load Index   (d) Index

Fig. 8. Box plots showing the questionnaire results conducted in 2025 ($n = 17$) encompassing the System Usability Scale (SUS) (a), statements on feedback quality (b) and the NASA Task Load Index (TLX) (c). The agreement in (a) and (b) ranges from *disagree at all* (--) and *rather disagree* (-) via *partly* (∅) to *rather agree* (+) and *agree at all* (++). The task load (d) is measured on a scale from *very low* to *very high*, except for *own performance* ranging from *perfect success* to *failure*. The aggregated raw SUS and TLX (d) range from 0 to 100.

we aggregate both the SUS and TLX without performing any prioritization or weighting. Hence, Fig. 8d only depicts the raw SUS and raw TLX. As can be seen, ASIA's usability scale is above average with a median 62.5, whereas the workload score for proving with ASIA is acceptable with a median of 40.

Discussion. We evaluated ASIA's suitability (RQ3) by investigating the number of automatically assessed and manually reassessed submissions. Our results (Fig. 5) indicate that the number of reassessments required is overall decreasing. We argue, however, that this is not only due to improvements in ASIA but rather stems from improvements made in teaching ASIA. For instance, from the year 2024 on, we provided a Wiki outlining its syntax and added more training exercises where students could immediately get feedback on their proofs. Moreover, we found that students who did the training exercises were able to provide syntactically correct proofs, which ASIA could assess automatically. Furthermore, from evaluating the automatically awarded points, in Fig. 7b, we conclude that ASIA fairly assesses the students' ability to employ structural inductions, in particular, when considering more difficult assignments, i.e., *Sets* and *Complex*. Thus, while we believe that ASIA could be employed without manual reassessments, if assignments are mandatory, award student credits, or are part of an exam, we maintain that manual reassessments will be necessary.

This is supported by our results shown in Fig. 7a, where grades were almost always improved by the reassessment. We attribute this to the reassessment, fixing minor syntax errors in a student's submission that inevitably happen in a stressful situation. Based on our experience and the evaluation results, we conclude that ASIA's suitability generalizes to larger courses. While the syntax and semantics of ADTs and structural induction proofs must be taught first, the benefits of an automatic assessment of students' submissions and immediate feedback for students far outweigh the reassessment effort. Especially in courses where context-free grammars and formal models with syntax and semantics are taught.

When considering the usability of ASIA (RQ4), we cannot evaluate it in isolation. Although ASIA is a command-line tool, our students use YAPEX to interface with ASIA. Hence, the results of the SUS are influenced by the usability of the online platform. Our evaluation indicated that ASIA's usability is above average (with a raw SUS score of 62.5), however is still considered cumbersome to use by many participants. Individual feedback of questionnaire suggested syntax highlighting and highlighting of incorrect braces as improvements. These features, however, must be implemented in YAPEX. In contrast, YAPEX has negligible impact on the feedback quality (RQ1), as it only lists the output provided by ASIA. Thus, we consider the results shown in Fig. 8b to be representative. In sum, most students agreed with statements on good feedback, i.e., that the feedback is relevant, helpful, easy to understand, and concise. Notably, though, some participants perceived the feedback to be insufficient to explain the mistakes they made. Last but not least, we investigated whether the introduction of ASIA for student assignments would increase the students' workload (RQ4). As a result of the TLX survey, this is not the case. In fact, the median perceived workload is below average (with a raw TLX score of 40). Participants reported high mental demand as well as medium to high levels of effort and frustration (cf. Fig. 8c). We argue, however, that the high mental demand and effort is inherent to the assigned task of proving a statement with structural induction, especially, for undergraduate students. In the case of the high *frustration*, this might stem from participants struggling to find syntax errors in their proofs, as indicated by ASIA. Again, a better integration of ASIA into YAPEX could reduce this issue. In conclusion, our evaluation indicates that ASIA's usability is above average, yet could be improved with a better integration into the exercise platform, e.g., underlining syntax errors.

Although we did not evaluate ASIA's performance, from our experience with ASIA in YAPEX, we can say that ASIA scales well for the assignments we provided, as each solutions could be automatically assessed in around 300ms. However, a benchmark and more thorough analysis will be necessary to assess the scalabilty of ASIA.

Threats to Validity. We followed the guidelines of [13] regarding threats to validity. To ensure *external validity*, we guaranteed that all students in the evaluation attended the corresponding course and that the course material did not significantly change. Still, the introduction of more training exercises and a wiki influenced positively impacted our results. To mitigate threats to *internal validity*, we left both the five assignments and grading schemes unchanged. Because two of the three student assistants graded all three years, their experience might have influenced the result. Regardless, the rules when and how a reassessment was conducted remained unchanged. Moreover, we addressed the expected low number of participants by employing two standardized questionnaires; still, the results are only indicative and by no means conclusive. Conversely, we do not consider the use of the raw SUS and raw TLX as a threat to validity. As a technical limitation, in contrast to ASIA, YAPEX only accepts integers as points. Consequently, the evaluation is limited to the decimal place of the corresponding floating point assessment. Similarly, the fact that ASIA is integrated into YAPEX might have impacted both the SUS and TLX results. To limit this influence, when evaluating the feedback quality, we explicitly asked students to rate the output of ASIA displayed verbatim. In case of *construct validity*, both the evaluation and questionnaire, were designed to answer the corresponding research questions. We did neither remove nor weigh questions to not include any biases. While the evaluation of the feedback quality did not employ an existing questionnaire, we maintain its adequacy, as it relies on existing guidelines for error messages [16] as well as a 5-point Likert scale, both established in their respective fields. To ensure *repeatability*, we provide a replication package [22] containing ASIA, the exercises, collected data, and the questionnaire.

8   Conclusions and Future Work

Since structural induction proofs are challenging for undergraduate students to learn, they require substantial feedback on their exercises. This makes a high effort for manual assessors. Thus, manual feedback and assessment is not scalable without engaging multiple assessors, which assess solutions individually, making the assessment and feedback quality potentially unfair. Thus, we need one single assessor, able to assess all undergraduate students and give them feedback. Due to the high effort, this can only be done by a machine, thus we need an automatic tool providing immediate feedback and assessment of structural induction.

We created an approach for proof checking tailored to generation of comprehensible feedback by considering follow-up errors and checking every mistake on its smallest context in the proof. Furthermore, we created a fair assessment schema for structural induction proofs, assigned to students. It is fair, as different students get the same assessment if they have the same mistakes. The feedback is comprehensible, as each mistake and the assessment for each subtask are presented to the student. We created a system which brings the theory of our proof checking approach and assessment schema into practice. We evaluated that system in exercises of our undergraduate course. As a result, the system has proven to be practical in real-world scenarios, and the assessment schema is fair, graduated, and scalable. The students found the feedback helpful and comprehensible. They also found that ASIA is practical to use and does not increase the task load compared to structural induction proofs on paper.

In the future, the assessment schema could be extended by nested structural induction or implications, conjunctions, and disjunctions of equations in proof goals. Moreover, we also want to evaluate the assessment of proofs with lemmas, which ASIA supports, yet was omitted for brevity.

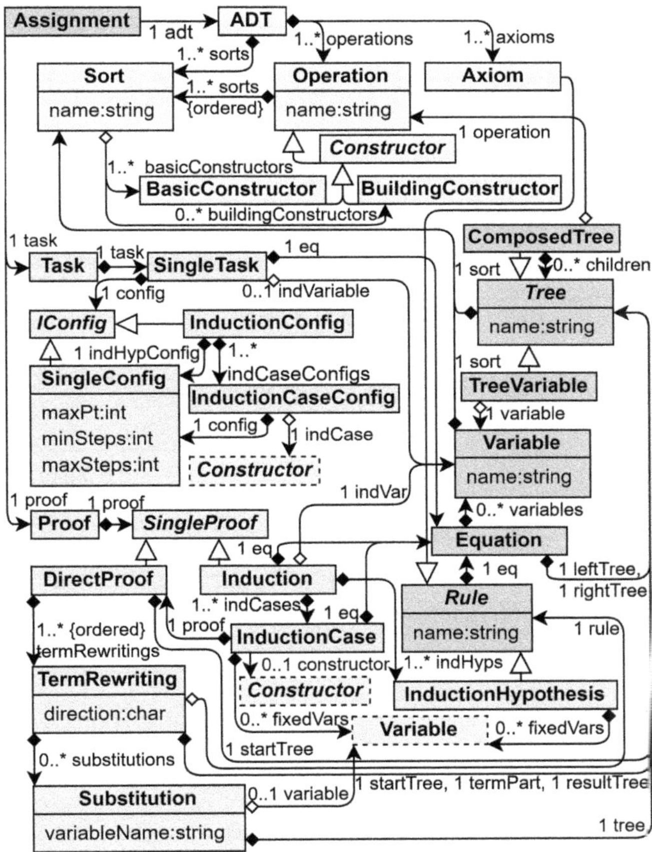

Fig. 9. Metamodel of the ADT (yellow), Task (green), Proof (orange), and the basic concepts they are using (light blue). (Color figure online)

A   Metamodel of Our Representation of ADTs, Tasks, and Proofs

We modeled the assignment (c.f. Fig. 3) in a class diagram. It consists of three main parts: ADT, Task and Proof, combined in the Assignment class (see Fig. 9, in blue). We omit the OCL constraints for brevity.

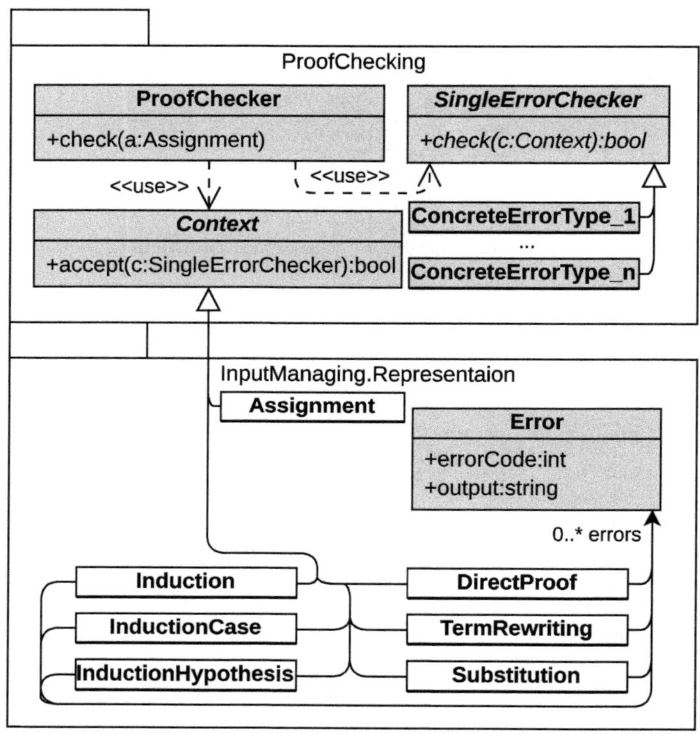

Fig. 10. Metamodel, extending Fig. 9 with concepts for checking proofs and error reporting.

For the proof checking, we first need to add the possibility in the metamodel to mark errors in the proof's representation. Therefore, the class Error is added (see Fig. 10). Each part of the proof, where errors can occur, has a list of errors. The error has a code for identification and an output string, which can contain additional information about the falsity, usually used for hints.

The proof checking implements a simplified version of the visitor design pattern: Each part of the proof where errors can occur and the assignment itself are contexts. The context can accept a single error checker to check that error on it. Then, we have the SingleErrorChecker class, which can check the context. Every concrete error type inherits from the single error checker. Thus, the single error checkers are visitors of the contexts. Since every checker applies to only one context, the model simplifies to only one check method for each `SingleErrorChecker`. That also makes the model extensible to new contexts, in contrast to the visitor design pattern. The check method of the `ProofChecker` class organizes the proof checking by assigning the checker to the contexts.

References

1. Autexier, S., Dietrich, D., Schiller, M.: Towards an intelligent tutor for mathematical proofs. Electron. Proc. Theoret. Comput. Sci. **79**, 1–28 (2012). https://doi.org/10.4204/eptcs.79.1
2. Back, R.J.: Structured derivations: a unified proof style for teaching mathematics. Formal Aspects Comput. **22**, 629–661 (2010). https://doi.org/10.1007/s00165-009-0136-5
3. Bertot, Y., Castéran, P.: Interactive Theorem Proving and Program Development: Coq'Art: The Calculus of Inductive Constructions. Springer, Heidelberg (2004). https://doi.org/10.1007/978-3-662-07964-5
4. Brooke, J.: SUS: a quick and dirty usability scale. In: Usability Evaluation in Industry, vol. 189, no. 194, pp. 4–7 (1996)
5. Burstall, R.M.: Proving properties of programs by structural induction. Comput. J. **12**(1), 41–48 (1969). https://doi.org/10.1093/comjnl/12.1.41
6. Conejo, R., Guzmán, E., Trella, M.: The SIETTE automatic assessment environment. Int. J. Artif. Intell. Educ. **26**, 270–292 (2016)
7. Dähne, J., Schiele, S., Thüring, A.: YAPEX: Yet Another Practical EXercise platform (2016). https://yapex.informatik.uni-halle.de/about/about_en.html. Accessed 16 Feb 2026
8. Hart, S.G., Staveland, L.E.: Development of NASA-TLX (task load index): results of empirical and theoretical research. In: Hancock, P.A., Meshkati, N. (eds.) Human Mental Workload, Advances in Psychology, vol. 52, pp. 139–183. North-Holland (1988). https://doi.org/10.1016/S0166-4115(08)62386-9
9. Ihantola, P., Ahoniemi, T., Karavirta, V., Seppälä, O.: Review of recent systems for automatic assessment of programming assignments. In: Proceedings of the 10th Koli Calling International Conference on Computing Education Research, Koli Calling '10, pp. 86–93. Association for Computing Machinery, New York, NY, USA (2010). https://doi.org/10.1145/1930464.1930480
10. Insa, D., Silva, J.: Automatic assessment of Java code. Comput. Lang. Syst. Struct. **53**, 59–72 (2018). https://doi.org/10.1016/j.cl.2018.01.004
11. International Organization for Standardization: ISO/IEC 14977:1996 Information Technology - Syntactic Metalanguage - Extended BNF (1996). https://www.iso.org/standard/26153.html
12. Kahl, W.: Calccheck: a proof checker for teaching the "logical approach to discrete math". In: Avigad, J., Mahboubi, A. (eds.) Interactive Theorem Proving, pp. 324–341. Springer, Cham (2018). https://doi.org/10.1007/978-3-319-94821-8_19
13. Konersmann, M., et al.: Evaluation methods and replicability of software architecture research objects. In: 2022 IEEE 19th International Conference on Software Architecture (ICSA), pp. 157–168. Institute of Electrical and Electronics Engineers, New York, NY, USA (2022). https://doi.org/10.1109/ICSA53651.2022.00023
14. Lodder, J., Heeren, B., Jeuring, J.: Providing hints, next steps and feedback in a tutoring system for structural induction. Electron. Proc. Theoret. Comput. Sci. **313**, 17–34 (2020). https://doi.org/10.4204/eptcs.313.2

15. Minh, F.T., Gonnord, L., Narboux, J.: A lean-based language for teaching proof in high school. In: de Paiva, V., Koepke, P. (eds.) Intelligent Computer Mathematics, pp. 447–467. Springer, Cham (2026). https://doi.org/10.1007/978-3-032-07021-0_25

16. Molich, R., Siehl, J., Schreyer, B., Lewandowska, D., Freund, E., Töpper, H.: Guideline for error messages. Technical report, German UPA e.V. (2024). https://germanupa.de/sites/default/files/2024-10/leitfaden-fehlermeldungen-v1.02-en_0.pdf. Accessed 29 Sept 2025

17. Moreno-León, J., Román-González, M., Harteveld, C., Robles, G.: On the automatic assessment of computational thinking skills: a comparison with human experts. In: Proceedings of the 2017 CHI Conference Extended Abstracts on Human Factors in Computing Systems, CHI EA '17, pp. 2788–2795. Association for Computing Machinery, New York, NY, USA (2017). https://doi.org/10.1145/3027063.3053216

18. Nipkow, T., Paulson, L.C., Wenzel, M.: Isabelle/HOL. A Proof Assistant for Higher-Order Logic, vol. 2283. Springer, Berlin (2002). https://doi.org/10.1007/3-540-45949-9

19. Owre, S., Rushby, J.M., Shankar, N.: PVS: a prototype verification system. In: Kapur, D. (ed.) CADE 1992. LNCS, vol. 607, pp. 748–752. Springer, Heidelberg (1992). https://doi.org/10.1007/3-540-55602-8_217

20. Patel, M., et al.: LeanTutor: towards a verified ai mathematical proof tutor (2026). https://arxiv.org/abs/2506.08321

21. Rebholz, S., Zimmermann, M.: Applying computer-aided intelligent assessment in the context of mathematical induction. In: Pan, Z., Cheok, A.D., Müller, W., Iurgel, I., Petta, P., Urban, B. (eds.) Transactions on Edutainment X. LNCS, vol. 7775, pp. 191–201. Springer, Heidelberg (2013). https://doi.org/10.1007/978-3-642-37919-2_11

22. Sabinus, E., Kühn, T., Zimmermann, W.: Automatic assessment and feedback on undergraduates' structural induction proofs, December 2025. https://doi.org/10.5281/zenodo.17777102

23. Segal, J.: Learners' difficulties with induction proofs. Int. J. Math. Educ. Sci. Technol. **29**(2), 159–177 (1998). https://doi.org/10.1080/0020739980290201

24. Wirsing, M.: Algebraic Specification, pp. 675–788. MIT Press, Cambridge, MA, USA (1991)

25. Zhao, C., Silva, M., Poulsen, S.: Autograding mathematical induction proofs with natural language processing. Int. J. Artif. Intell. Educ. **35**(5), 3202–3232 (2025). https://doi.org/10.1007/s40593-025-00498-2

Seven-Year Activity to Introduce B-Method to Japanese Technical College Students

Takaomi Ohnishi$^{(\boxtimes)}$ (iD), Yoshihiko Nakamura (iD), and Ryota Yamamoto (iD)

National Institute of Technology, Tomakomai College, Tomakomai, Hokkaido, Japan
{ohnishi,ynakamura,r-yamamoto}@tomakomai-ct.ac.jp

Abstract. This article reports our 7-year activity to introduce the formal method, B-Method to our college. Our activity mainly includes the mandatory short-term experiment for the whole member of the 4th-grade class (the age of 18 or 19), with few prerequisites about the logical mathematics. Under the time constraints - the course of only 140 min. per week, two or three weeks -, our experiment covers all of three fields: formal specification, proof obligation generation and model checking, and also gives the students the opportunities to briefly experience of state explosion problem. The authors set the educational goal for the experiments as giving opportunities to commit a formal method as the software engineer's fundamental knowledge. The authors also arranged the short-time - at most 3-hour - lectures for people other than the students mentioned above who cannot write a logical predicate at all. The goal of this lecture is to give them the opportunity to find tacit facts with sound reasoning from documents. On a web page (https://www.asahi-net.or.jp/~ny9t-oons/b-method_english.html), the authors uploaded details - course materials, contents of models and the results of survey questions -. As far as the data of the survey questions show, it seems that the educational goal for the experiments is achieved, our experiment and lecture seem to be generally accepted, but there are some abilities to improve the methodology within the time constraint of the course.

Keywords: formal methods · B-Method · student experiments · mathematical reasoning

1 Introduction

In software system development projects, specifications written in natural language often contain ambiguous expressions. This ambiguity makes it difficult to identify contradictions or errors within the specification content. Formal methods are worth using to eliminate this ambiguity due to their strict specification descriptions. Technical education is necessary to promote and advance formal methods. It is said that the barriers to their introduction are high, there is an issue with developing training to make sure we have enough human resources. For overcoming this issue, the authors approach design the methodology to teach a formal method to young generations.

2   Our Activity Overview - Our College Student's Experiment

2.1   Our College Student's Experiment - Overview

The authors have designed and launched experiments with students using the formal method, B-Method applied for the whole member (at most 40 persons) of the 4th-grade class (the age of 18 or 19) of our college - National Institute of Technology, Tomakomai College - [1]. Each experiment is 140 min per week. They are held for two weeks from 2019 to 2024 and are expanded to three weeks in 2025. Students are members of Computer Science and Engineering Course, who have essential coding skill of the C language, Java, and Python, but have no knowledge nor experience about formal method, few knowledge nor experience about logical mathematics.

Under the time constraints of the experiments, the authors set the educational goal for the experiments as giving opportunities to commit a formal method as the software engineer's fundamental knowledge. Due to this educational goal, the authors designed the experiments as including all of formal specification, proof obligation generation, and model checking. The formal method, B-Method has an aptitude for this request. Due to the student's capability as a beginner of formal methods, the authors apply Atelier B [2,3] for formal specification and proof obligation generation, and ProB [4,5] for model checking with the usage of LTL expressions.

The teachers of the experiments studied with some educational materials [6,7] to acquire the skill of teaching B-Method to the students.

2.2   Our College Student's Experiment - Week 1

As the beginning week of the student's experiment, the authors have applied deductive logic puzzles [8,9] (Fig. 1 is a sample, modified by the authors due to copyright) as the "static model," that consists of only the one proper state - correct answer of the puzzle - . The teachers of the experiments provide SETS clause and CONSTANTS clause of the B-Method model (Fig. 2) except for the predicates of PROPERTIES clause. Students read the puzzle - both the main sentences which some tacit facts may be hidden (as the example of Fig. 1, the person aged b is not X, and etc.) and the constraints - edit predicates from them, write the predicates to ProB on PROPERTIES clause of the B-Method model. And finally, reload the formal model of ProB in order to activate the written predicates of PROPERTIES clause, and observe the reduction of solution space. This is the student's work (Fig. 3).

2.3   Our College Student's Experiment - Week 2

As the second week of the student's experiment, the authors have applied the questions from the test of the lecture "logic circuit" (Fig. 4), that consists of at most three proper states - capable scale for students -, as a "tiny dynamic

There are four persons: W, X, Y, and Z,
who are all of different ages: a, b, c, and d, not respectively.
Each of these owns one or more items.
•The person aged b and X together own a total of four items.
•The person aged c and Y together own a total of three items.
•The person aged a and Z together own a total of three items.
•The person aged c and W together own a total of four items.
•Four persons W, X, Y and Z together own a total of eight items.
Describe in detail the relationship between the people, their ages,
and the number of items they own.
(Answer: W-d-3, X-a-2, Y-b-2 and Z-c-1.)

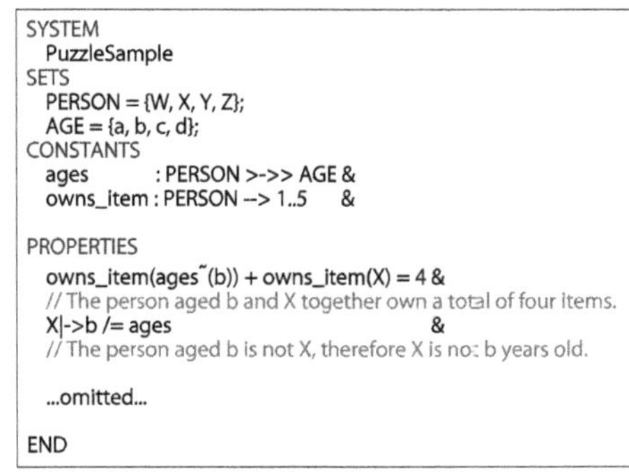

```
SYSTEM
  PuzzleSample
SETS
  PERSON = {W, X, Y, Z};
  AGE = {a, b, c, d};
CONSTANTS
  ages        : PERSON >->> AGE &
  owns_item : PERSON --> 1..5    &

PROPERTIES
  owns_item(ages˜(b)) + owns_item(X) = 4 &
  // The person aged b and X together own a total of four items.
  X|->b /= ages                          &
  // The person aged b is not X, therefore X is not b years old.

  ...omitted...

END
```

Fig. 1. Summary of both question and answer of a deductive logic puzzle

Fig. 2. Semantic mapping about Fig. 1 logic puzzle onto the syntax structure of B-Method

model." As far as Fig. 4 - there are $State Q_1 Q_0$ with four states, $Input X_1 X_0$, and $Output Z_1 Z_0$ -, this is a question to derive the logical expressions of a sequential logic circuit using Karnaugh maps. State 00, 10 and 11 are proper states, and state 01 is an improper state. Our goal of experiment Week 2 is to make a formal model that never transits to the improper state 01, so "three proper states." In a state spaces with such few three states, it is difficult to establish a common understanding between a teacher and students. (Why state $Q_1 Q_0 = 01$ is defined as improper? People out of the lecture "logic circuit" class may not guess it.) The lecture "logic circuit" is the student's previous year's one, thus, the test of the lecture is so familiar to them.

Figure 5 shows the concept of experiment Week 2. First, the teacher gives the students a question of the test of lecture "logic circuit," and shows the definitions about state values, state variables, and operations to edit B-Method specification. At most two state variables - each of them has two state values -, and four operations. Second, Students edit four state transition tables for each operation. Third, students edit the formal specification as an abstract machine component of Atelier B. In editing the abstract machine component, the teacher teaches students to use the way of "becomes such that substitutions" - that would be a template from the state transition table - to edit each operation. Fourth, students try to generate proof obligations and prove them (automatic and interactive) of the abstract machine component. Interactive proving will be at most pressing PP1 button. Simultaneously, students try the model checking with the usage of LTL expressions. Fifth, students refine the abstract machine component to an implementation component. In refinement to the implementation component, the teacher teaches students to use the way of "IF conditional substitutions" - which is so familiar to the students - to edit each operation. Sixth, students try the proof obligation generation, proving them, and model checking to the implementation component. And finally, students command to generate code of the C language from the implementation component.

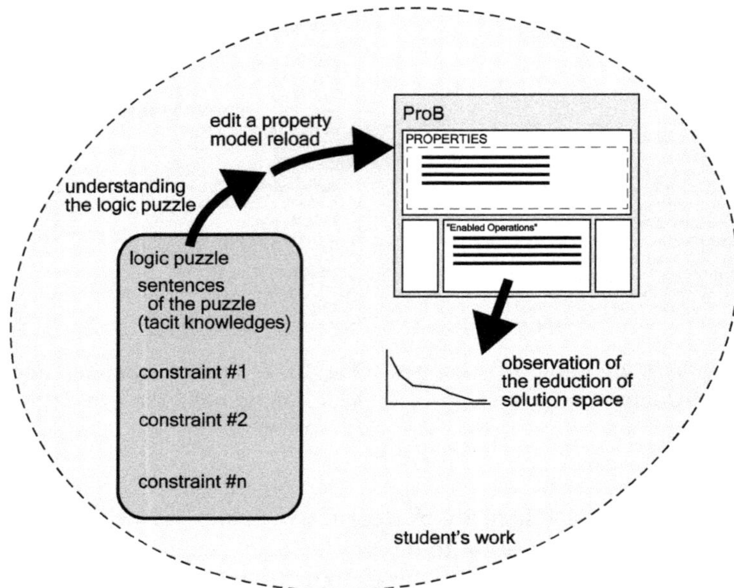

Fig. 3. The concept of experiment Week 1

問　　りょうた君は彼女のあゆみちゃんがなわ飛びをするのを見るのが好きです。
　　　あゆみちゃんは踊りながらロープをひろうことはできません。
　　　表1のように論理変数を割り当てたときに、JKフリップフロップを用いて
　　　順序論理回路を設計して、必要な論理式を全て答えよ。
　　《Mr. Ryota likes to see her girl friend, Ms. Ayumi, jumping rope. She cannot pick any
　　rope with her dancing. Design a sequential logic circuit using JK flip-flops. Answer
　　all of the essential boolean expressions with the assumption that the boolean variables
　　are assigned as "表 1 ".》

表 1　順序論理回路の変数の割り当て

論理変数の 名称	論理変数が持つ真理値の意味
状態： Q_1Q_0	**00** : 何も持たずに、だまって立っている。 《With empty hands, she is just standing.》 **01** : 何も持たずに、踊っている。 《With empty hands, she is just dancing.》 **10** : ロープを両手に持ったまま、だまって立っている。 《Holding a rope with her both hands, she is just standing.》 **11** : ロープを両手で回して、なわ飛びをする。 《Turning a rope by her both hands, she does jumping rope.》
入力： X_1X_0	**00** : ロープを捨てる、あるいはロープをひろおうとする。 《Throw the rope away, or attempt to pick the rope up.》 **01** : 体を動かす。《Move her body.》 **10** : 体を動かすのを止める。《Stop moving her body.》 **11** :（ドントケア。）《(Don't care.)》
出力： Z_1Z_0	**00** : なわを飛ぶことが期待される。《Hoped to jump the rope.》 **01** : なわ飛びの成功をよろこぶ。 《Being glad to see the success of her jumping rope.》 **10** : 踊りを止めることが期待される。《Hoped to stop dancing.》 **11** : ロープを拾うことが期待される。 《Hoped to pick the rope up.》

Fig. 4. A question from the test of the lecture "logic circuit"

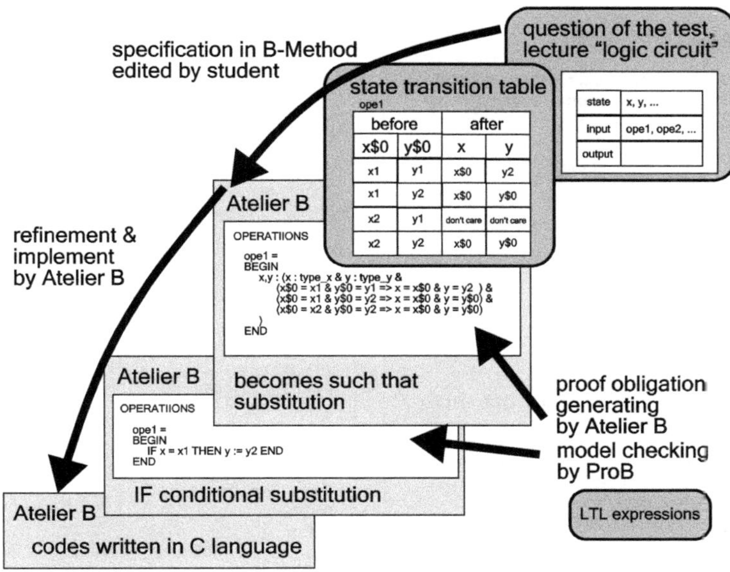

Fig. 5. The concept of experiment Week 2

In the case of Fig. 4, the teacher defined operations as "pick_rope_up," "throw_rope_away," "move_body," and "stop_body" newly arranged from the row "$InputX_1X_0$" of the table of Fig. 4. Figure 6 shows the state transition table for the operation "throw_rope_away," Fig. 7 shows the sample of the abstract machine component, and Fig. 8 shows the sample of the implementation component.

Operation : throw_rope_away

before		after	
rope$0	girl$0	rope	girl
rope_not_holding	just_standing	rope$0	girl$0
rope_not_holding	rope_jumping	(don't care)	(don't care)
rope_holding	just_standing	rope_not_holding	g rl$0
rope_holding	rope_jumping	rope$0	g rl$0

Fig. 6. Sample of a state transition table for the operation throw_rope_away of the lecture "logic circuit" test question Fig. 4

2.4   Our College Student's Experiment - Week 3

The authors launched Week 3 in 2025 as the final week of the student's experiment, the authors have introduced both the visual interface by ProB and the

```
MACHINE
  JumpingRope
SETS
  type_rope = {rope_not_holding, rope_holding};
  type_girl = {just_standing, rope_jumping}
ABSTRACT_VARIABLES
  rope, girl
INVARIANT
  rope : type_rope & girl : type_girl &
  (rope = rope_not_holding => girl = just_standing) &
  (girl = rope_jumping    => rope = rope_holding )
INITIALISATION
  rope := rope_not_holding ||
  girl := just_standing
OPERATIONS
  pick_rope_up = BEGIN   ...omitted...  END;

  throw_rope_away =
  BEGIN
    rope, girl : (rope : type_rope & girl : type_girl &
    (rope$0 = rope_not_holding =>
      rope = rope$0 &
      girl = girl$0                ) &
    (rope$0 = rope_holding      =>
       (girl$0 = just_standing =>
          rope = rope_not_holding &
          girl = girl$0           ) &
       (girl$0 = rope_jumping =>
          rope = rope$0 &
          girl = girl$0            ) )        )
  END;

  move_body = BEGIN   ...omitted...  END;
  stop_body = BEGIN   ...omitted...  END

END
```

Fig. 7. Sample of an abstract machine component

```
IMPLEMENTATION
  JumpingRope__i
REFINES
  JumpingRope
CONCRETE_VARIABLES
  rope_i, girl_i
INVARIANT
  rope_i : BOOL & girl_i : BOOL &
  ((rope_i = FALSE) <=> (rope = rope_not_holding)) &
  ((rope_i = TRUE ) <=> (rope = rope_holding     )) &
  ((girl_i = FALSE ) <=> (girl = just_standing    )) &
  ((girl_i = TRUE  ) <=> (girl = rope_jumping     ))
INITIALISATION
  rope_i := FALSE;
  girl_i := FALSE
OPERATIONS
  pick_rope_up = BEGIN   ...omitted...  END;

  throw_rope_away =
  BEGIN
    IF
       rope_i = TRUE & girl_i = FALSE
    THEN
       rope_i := FALSE
    END
  END;

  move_body = BEGIN   ...omitted...  END;
  stop_body = BEGIN   ...omitted...  END

END
```

Fig. 8. Sample of an implementation component

"error capable model" concept mentioned in section3.2. The authors applied the "vending machine" model from the text of model checking [10] and enhanced some functions into the model. Both repairing the broken (by the teacher) model and reporting about the repair process are student's work. Our "vending machine" model has 6 state variables, 2 output variables, and 12 operations. The authors edited the specification written in natural language of 28 items, and 47 LTL expressions (13 correspond to the model checking text mentioned above, 34 are made by the authors, some overlap) to be asserted. Some of LTL expressions are made with the usage of Before-After temporal logic operator (BA), that is available since ProB 1.12 [4]. Figure 9 is the face of the visual interface of the "vending machine" model - the yellow (pale color) area is the front side for customers, the black (dark color) area is inside the maintenance door for the provider, and a star exists on the upper right is the reset button of the "error capable model" -. And Fig. 10 is at the locked "error" state by the broken operation "coin_insert", when the coin insertion block is clicked, then both the coin insertion block and the reset button turn red. When a student clicks the reset button, the system "vending machine" is unlocked and returns to its initial state.

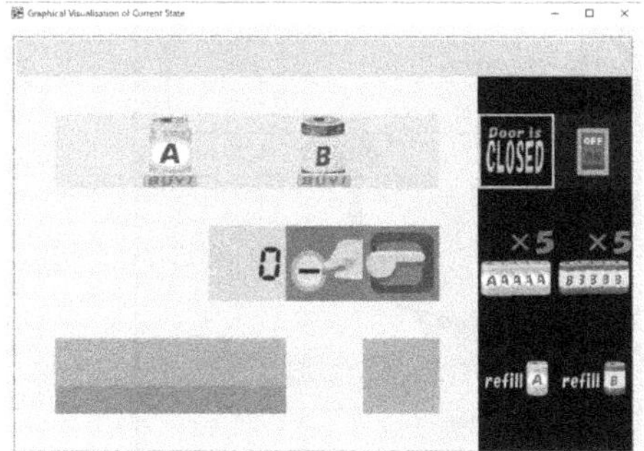

Fig. 9. "vending machine" model

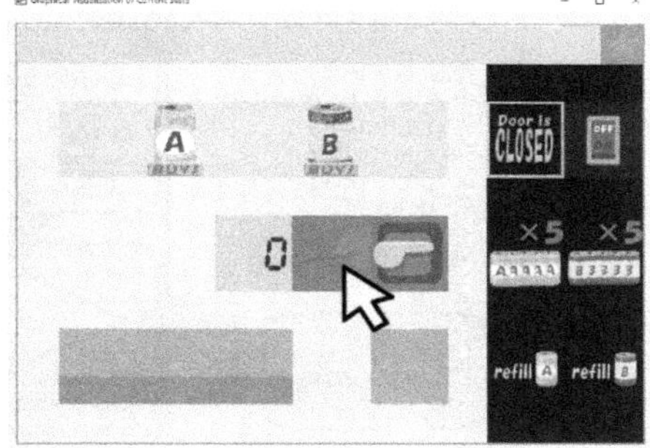

Fig. 10. "vending machine" model at the locked "error" state

Figure 11 shows a structure of the "vending machine" models with both the visual interface by ProB and the "error capable model" concept. Students face only two non-colored blocks of this figure - the "vending machine" model to be repaired by the students and the visual interface by ProB -. Three deep-colored blocks - asserted LTL expressions, correct model for verify, and the model for comparison - should be hidden from students. A pale-colored block is the model, that includes all of the models mentioned above and runs the visual interface. Due to the restriction on ProB, both the model to be repaired by the students and the correct model for verify must be in a same directory, and due to the restriction of the circumstance of the experience, the correct model for verify is readable by students no matter what. The authors have arranged that the cheating actions would be wasted efforts for the students. For example, the correct model just satisfies only the post conditions of the operations, ignoring preconditions, uses the way of substitution different than the model to be repaired by the students. Due to the larger state spaces of the "vending machine" models than the model of Week 2, the authors arranged the students to briefly experience state explosion problem - which the proper formal methods engineers will eventually face - within the time restriction of the experiment.

2.5   Results of the Experiment

Due to the mandatory experiment, all students achieved all tasks of all Week, and submit the reports and the answer of survey questions.

In Week 1, the students made the static models which solve the deductive logic puzzles to correct answers. In Week 2, the students made the state transition tables from the question of the test, made the tiny dynamic models from the state transition tables, and generated proof obligations, proved them, and

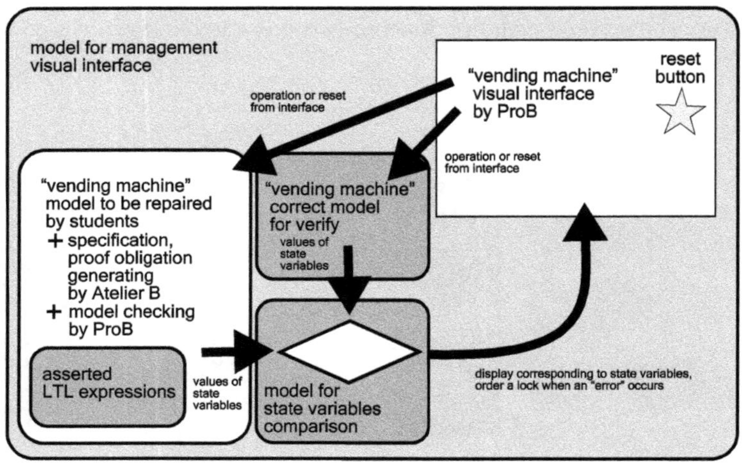

Fig. 11. The concept of experiment Week 3

model checked the models. The students refined and implemented the models to generate the codes written in the C language. In Week 3, including subjective insights, it seems that the students did not feel the complexity gap - size of formal models - compared to Week 1 and Week 2. (The authors would like to believe that's the result of using "Error Capable Model"(section3.2.)) Table 1 is the summary of the results of Week 3. The detailed results are shown on the web page, and the URL is shown in the abstract on the top page of this article. The authors actually introduced 8 defects into the student's model this year. The difference in the result does not indicate the difference of ability between the proof obligation generation and the model checking. Due to the state explosion problems in early model checking by ProB, the teacher has instructed the students to generate proof obligations by Atelier B first.

Table 1. Results of Week 3: 2025's student's defect locations detection of "vending machine" model

detected defect locations	N	ave.	max.	$Q_{3/4}$	$Q_{2/4}$	$Q_{1/4}$	min.
all of defect locations	34	8.03	10	9	8	8	6
detection by ProB animation interface	34	7.15	10	8	8	7	1
detection by Atelier B proof obligation generation	34	3.38	8	5	4	0	0
detection by ProB LTL expression proving	34	1.68	9	3	1	0	0

3   Our Activity Overview - Serendipitous Ideas of Our Activity

3.1   Fountain of Logic Expressions - Short-Time Lectures to the Other Than the Experiment Members

Let us assume that kids who are elementary school or junior high school students, or college students who are as young as high school students, or who are other than computer science nor engineering or are not Japanese domestic. Naturally, they cannot write B-Method predicates at all. The authors have arranged short-time lectures for them with the goal of giving them the opportunity to find tacit facts with sound reasoning from documents for introducing mathematical reasoning [11].

To simplify this topic, let us assume the case of three-hour lecture to kids (Fig. 12).

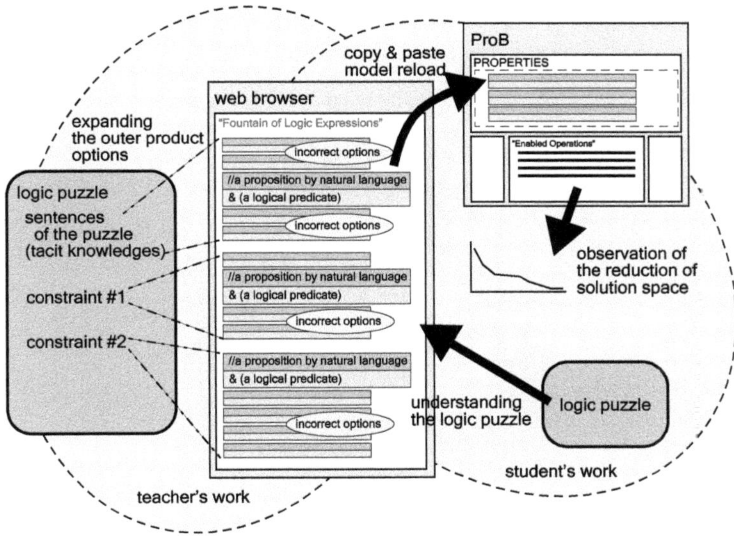

Fig. 12. The concept of kid's short-time lecture

Teacher makes a web page of "Fountain of Logic Expressions." This "Fountain of Logic Expressions" is an aggregation of the pairs of a logical predicate and a proposition by natural language. All of the logical predicate are written in B-Method, which kids probably cannot understand. All of the propositions by natural language are readable by kids, are commented out not to be parsed as B-Method. First, the teacher picks up all of the puzzle elements - tacit knowledge hidden within the main sentences, and the constraints -. Second, expands each of the elements toward outer product options. So, most of the options are incorrect for the puzzle. Third, edits the predicate, which corresponds to the option. And last, makes a pair of the predicate and the proposition, this pair is the part of the "Fountain of Logic Expressions." This is the teacher's work.

The kids, who attend to the lecture, first, read and understand the meaning of logic puzzle, but are taught not to solve the puzzle by themselves. Second, choose a pair - which they believe to be correct for the puzzle - of a logical predicate and a proposition by natural language, from the web page of "Fountain of Logic Expressions." Third, copy the chosen pair from the web browser and paste it as a property of ProB. And finally, reload the formal model of ProB in order to activate the predicate of the chosen pair as a property, and observe the reduction of solution space. This is the kid's work.

The authors also applied this idea of "Fountain of Logic Expressions" to the puzzle game "Sokoban (warehouse keeper) [12]" as shown in Fig. 13. The "Sokoban" is other than the "static" logic puzzles, in the meaning of applying "Fountain of Logic Expressions" toward "dynamic" model. This trial for the kid's short-time lecture leads the authors to an idea of experiment Week 3 mentioned in section2.4 and section3.2.

For the evaluation of the short-time lectures mentioned above, the authors conducted easy questionnaire surveys for the members.

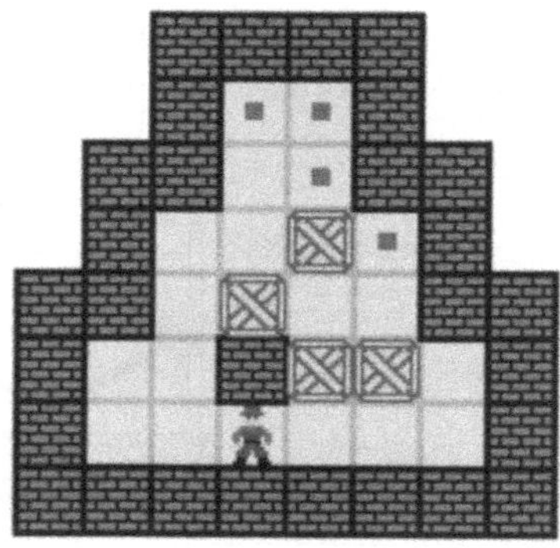

Fig. 13. the puzzle game "sokoban"

3.2 Error Capable Model - Concept for Student's Experiment Week 3

For overcoming the educational weak points of Week 2 experiment (section2.3) - found from the results of student's questionnaire (section2.5) -, the authors apply "error capable model" concept to formal model.

Figure 14 is an image of the state transition chart of systems made by formal methods. Of course, the boundary - either the safety is guaranteed or not - is rigidly defined by logical predicates. When students build safe systems and verify systems in existing ways, they have to look at a lot of information only by characters - logical expressions, values of variables -. (Whenever changing to green screens makes them so delightful.) For beginners like our college students, this situation, especially in the Week 2 experiment, makes the student's evaluation about the familiarity to formal methods decrease.

So, the authors have arranged the image of state transition chart of systems toward the "error capable model" shown in Fig. 15. First, the authors choose an initial state. Second, defines the states enveloped by a single layer out of the first boundary as "error" states, and also defines the second boundary that encloses the "error" states. And finally, sets a reset button beside the system. Any transition to any of the "error" states is capable, but whenever the state once transits to any of the "error" states, the state transition will immediately be locked. Resetting to the initial state of the system by pressing the reset button is always acceptable, regardless of whether the current state transition is locked or unlocked.

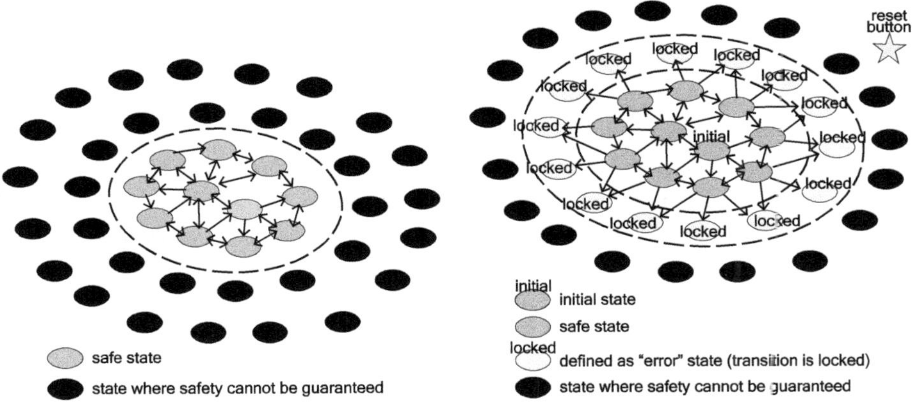

Fig. 14. The state space of the normal systems made by formal methods

Fig. 15. The state space of the systems made with the concept "error capable model"

4 Evaluation for Our Activity

4.1 Students' Questionnaire Survey

Under time constraints of the experiments, the authors did not conduct the tests for the students but the weekly questionnaire surveys. The authors have arranged the survey questions by referring to a well-known criteria [13] - Level-0 though Level-2 of formal methods usage - which remain an article of Japanese Wikipedia [14] to this day. Referring to the Japanese authority's commentary

[15] about these Levels, the authors have refined the Levels as shown in Table 2. The authors have also made learning goals of experiment as shown in Table 3, and made the relationship between the refined Levels and the learning goals as shown in Table 4. All survey questions - five-grade evaluation of 5 though 1 - for the students are made corresponding to each of the learning goals, as a part is shown in Table 3. The whole survey questions and detailed results are shown on the web page, and the URL is shown in the abstract on the top page of this article.

Table 5 is the data of the averages and quartiles of the results of Week 2's student's survey questions. (Week 2 is much harder for students than Week-1 due to the increased task under time constraints. The data for Week 3 are too few to tell the trend because of only one trial.) The averages are in the high 3 s or more. The medians $Q_{2/4}$ are 4 or 5. The inter-quartile ranges ($Q_{3/4} - Q_{1/4}$) are 1 or 2.

Table 2. Refinement for each the Level of formal methods usage

Level-0	Formal specification may be undertaken and then a program developed from this informally.
[L0-1]	having ability to describe the rigid specification with the usage of mathematical notation.
[L0-2]	having ability to develop the software by referring to the rigid specification edited in mathematical notation.
Level-1	Formal development and formal verification may be used to produce a program in a more formal manner.
[L1-1]	having ability to prove the quality of the software
[L1-2]	having ability to develop the software from the specification by the manner of refinement
Level-2	Theorem provers may be used to undertake fully formal machine-checked proofs.
[L2-1]	having ability to prove the quality of the software with the usage of proof obligation generator
[L2-2]	having ability to verify the quality of the software with the usage of model checker

4.2   Comparison with Other Educational Activities

Outside of Japan, upon referring to the thought-provoking FMTea papers, the activity of UFPE, Brazil [16], is closely related to ours - educates B-Method, covers all of three fields: formal specification, proof obligation generation and model checking -. The students need prerequisite lecture of software engineering.

Table 3. Week 2's learning goals and survey questions

About Atelier B	
	(2-A) through (2-F) .... omitted ....
About ProB	
	(2-G) through (2-K) .... omitted ....
About the whole of B-Method	
(2-L)	having ability to describe the formal specification with logical expressions from natural language
	Do you think you become used to B-Method logical expressions?
(2-M)	having ability to understand the significance of formal model checking's coverage to whole states of systems
	Do you think you were able to understand that you can improve the system's dependability by the usage of theorem proving and by the usage of checking by LTL expressions?
(2-N)	having ability to understand the significance that formal methods contribute to improve the system's dependability
	Do you think you were able to understand that you can verify the system has no defect by the usage of LTL expressions with checking whole states of the system?
(2-O)	having ability to understand the significance that formal specification is the proper approach to contribute to implement defect-free codes
	Do you think you understand that formal methods are useful to implement no-defect code in contrast to simply creating code that anyways works?
(2-P)	having ability to agree the significance of the process of formal specifications, theorem proving and formal model checking
	Do you think you feel familiar to formal method, B-Method through both example and exercises of the experiment?
(2-Q)	having the opportunity to commit a formal method as the software engineer's fundamental knowledge
	Do you think you were able to have the opportunity to commit a formal method as the software engineer's fundamental knowledge?

Table 4. Mapping from learning goals and survey questions, to the Levels of formal methods usage

Learning Goals & Survey Questions	Week 1	Week 2	Week 3
Level-0: Formal Specification			
[L0-1]	(1-A)(1-B) (1-C)(1-D) (1-E)(1-J) (1-K)(1-L)	(2-A)(2-B) (2-L)	(3-A)(3-B) (3-J)
[L0-2]	-	(2-C)(2-O)	(3-C)(3-M)
Level-1: Formal Development and Formal Verification			
[L1-1]	(1-E)(1-F) (1-G)(1-H)	(2-D)(2-E) (2-I)(2-J) (2-N)	(3-D)(3-H) (3-L)
[L1-2]	-	(2-C)	(3-C)
Level-2: Theorem Provers			
[L2-1]	(1-E)(1-M) (1-O)	(2-C)(2-D) (2-E)(2-F) (2-P)(2-Q)	(3-C)(3-D) (3-E)(3-N) (3-O)
[L2-2]	(1-F)(1-G) (1-H)(1-I) (1-M)(1-O)	(2-I)(2-J) (2-K)(2-P) (2-Q)	(3-H)(3-I) (3-N)(3-O)

The activities of both Exon, UK [17] and the course conducted at RU, Iceland [18] are also related to ours - education complete in 3 weeks -. It is worth to understand that each of them is accomplished under the fundamental knowledge of network security and real-time modeling, respectively.

The activities of Coder Dojo by SUT [19], Thailand and ANU, Australia [20] are related to ours in a different meaning - the approach to introduce a formal method for kids and students with few knowledge about logic mathematics -.

In Japan, the Top-SE project of NII [21] is the most famous. This ongoing project is provided for professionals and includes both formal specification and model checking. Looking into Japanese technical colleges like us - National Institute of Technology (NIT), Tomakomai College -, there are a couple of colleges that teach formal methods through lectures and experiments. As one of the most active cases in NIT, Numazu College conducts a half-year lecture - actually a 15-week lecture and tests, reports - on formal verification [22]. that is to as same grade students as our experiment. But in most technical colleges, formal methods are treated as at most a single topic of the software engineering lectures, moreover, most of those lectures are for advanced course - as same as second-half

Table 5. Results of Week 2's survey questions (five-grade evaluation of 5 through 1)

Survey Questions	N	ave.	max.	$Q_{3/4}$	$Q_{2/4}$	$Q_{1/4}$	min.	Survey Period
About Atelier B								
(2-A)	239	4.11	5	5	4	4	2	2019–2025
(2-B)	239	4.11	5	5	4	4	2	2019–2025
(2-C)	239	4.16	5	5	4	4	2	2019–2025
(2-D)	239	4.21	5	5	4	4	2	2019–2025
(2-E)	202	4.19	5	5	4	4	1	2020–2025
(2-F)	239	4.23	5	5	4	4	1	2019–2025
About ProB								
(2-G)	239	4.33	5	5	5	4	1	2019–2025
(2-H)	239	4.12	5	5	4	4	1	2019–2025
(2-I)	239	4.27	5	5	4	4	1	2019–2025
(2-J)	239	4.27	5	5	4	4	1	2019–2025
(2-K)	239	4.22	5	5	4	4	1	2019–2025
About the whole of B-Method								
(2-L)	239	3.79	5	4	4	3	1	2019–2025
(2-M)	238	3.90	5	5	4	3	1	2019–2025
(2-N)	239	4.13	5	5	4	3	1	2019–2025
(2-O)	238	4.39	5	5	5	4	2	2019–2025
(2-P)	240	4.08	5	5	4	3	1	2019–2025
(2-Q)	202	4.42	5	5	5	4	3	2020–2025

of undergraduate course for the age of 20 though 22 - students. The case of our activity - all-class-member's mandatory short-term experiments for young students with few prerequisites, ages as same as first-half of undergraduate course, covers all of three fields: formal specification, proof obligation generation and model checking - is original.

5 Discussion and Conclusion

5.1 Discussion

The authors have conducted the experiment Week 1 and Week 2 for 7 years. It includes subjective insights, Week 1 is accepted by the students. The medians $Q_{2/4}$ of the result of 14 survey questions are 4 or 5. The inter-quartile ranges $(Q_{3/4} - Q_{1/4})$ are 0 through 2 - most wide-distributed "2" was for only one question -. For the students, the logical predicate's effect of obtaining the correct answer puzzles is easy to understand.

In Week 2, the large number of tasks - due to the acquisition of the learning goals - within the time constraint, the students seem to be too tired to understand the essence of the learning goals. As mentioned above, the medians $Q_{2/4}$ are 4 or 5, the inter-quartile ranges $(Q_{3/4} - Q_{1/4})$ are 1 or 2. The data is not significantly different from Week 1, through the authors felt the need for an additional week.

In 2025, the additional experiment Week 3 was launched. The data is almost same with Week 2, the medians $Q_{2/4}$ are 4 or 5, the inter-quartile ranges $(Q_{3/4} - Q_{1/4})$ are 1 or 2. The authors feel that Week 3 provides the rationale for the Week 2's existence as the prerequisite for the students. Referring to Table 1, some of the students seem to be so simplistic that they are addictive on the animation interface - if the state transition is found locked, they just look for the failure in the model -. The authors need more clever educational methodology to utilize formal methods. The goal should be making the process that the students detect and identify defect locations and repair them with sound reasoning.

As mentioned above (section 2.1), the authors set the educational goal for the experiments as giving opportunities to commit a formal method as the software engineer's fundamental knowledge. The authors set the same final survey questions for the students on every week as "Do you think you were able to have the opportunity to commit a formal method as the software engineer's fundamental knowledge?" about the whole of B-Method for three weeks. Table 6 is the data of the averages and quartiles of the results of every week's student's survey questions. The averages are over 4, the medians $Q_{2/4}$ are 4 or 5, the inter-quartile ranges $(Q_{3/4} - Q_{1/4})$ are 1. It seems that the educational goal for the experiments is achieved.

As far as the data of the survey questions show, our experiment seems to be generally accepted, but there are some abilities to improve the methodology especially Week 3 within the time constraint.

The authors also acknowledge that our activity has evaluation shortcomings. By any chance, if the authors have ever expected that our activity would have the ability to train experts and specialists in formal methods, we evaluated the students with the usage of tests or long-term surveys. Naturally, and even now, the authors believe that this period is too short to develop the ability to train experts and specialists in formal methods. The authors are honestly surprised that our course has reached the step of repairing formal models within the time constraint. The authors understand that there is a discussion about our activity as is worth providing to the students the software engineer's fundamental knowledge "in long term." The authors set the educational goal for the experiments as "giving opportunities" to commit a formal method as the software engineer's fundamental knowledge. Anyway, the authors acknowledge and regret that we have never evaluated student's competence in long term.

Regarding the short-time lecture for people other than the students mentioned above, while there is an initial bias that the member expressed interest to join, they generally seem to accept the lecture.

Table 6. Results of every experiment week's final survey question for the students "Do you think you were able to have the opportunity to commit a formal method as the software engineer's fundamental knowledge?" about the whole of B-Method

week	N	ave.	max.	$Q_{3/4}$	$Q_{2/4}$	$Q_{1/4}$	min.	Survey Period
				Quartiles				
Week 1	202	4.46	5	5	5	4	2	2020-2025
Week 2	202	4.42	5	5	5	4	3	2020-2025
Week 3	36	4.28	5	5	4	4	1	2025

5.2 Conclusion

In this article, the authors reported the 7-year activity to introduce the formal method, B-Method to Japanese technical college.

The activity mainly includes the all-class-member's mandatory 3-week experiments for young students with few prerequisites, ages as same as first-half of undergraduate course, covers all of three fields: formal specification, proof obligation generation and model checking. In the experiment Week 1, the authors applied deductive logic puzzles as the "static model." In Week 2, the authors applied the questions from the test that consists of at most three proper states as a "tiny dynamic model." In Week 3, the authors introduced both the visual interface by ProB and the "error capable model" concept. The students repaired the broken "vending machine" model. The educational goal for the experiment is set as giving opportunities to commit a formal method as the software engineer's fundamental knowledge. For evaluation of the experiments, the authors conducted weekly questionnaire surveys for the students. The survey questions are arranged by referring to well-known criteria of formal methods usage.

The authors also arranged a short-time lecture for people other than the students mentioned above who cannot write a logical predicate at all. The authors introduced "Fountain of Logic Expressions" concept. The goal of this lecture is to give them the opportunity to find tacit facts with sound reasoning from documents for introducing mathematical reasoning. For evaluation of the lectures, the authors conducted easy questionnaire surveys for the members.

In this way, the simultaneous conducting both the experiment and the lecture mentioned above provided the improvement of our methodology of introducing B-Method to teach a formal method to young generations.

As far as the data of the survey questions show, it seems that the educational goal for the experiments is achieved, our experiment and lecture seem to be generally accepted, but there are some abilities to improve the methodology within the time constraint.

References

1. Ohnishi T., Yoshimura H., Abe T., Inagawa K., Yamamoto R., Hori T.: A practice of student experiments of the formal method, B-method applied for students of institute of technology. J. IPSJ **61**(4), 863–883 (2020). https://doi.org/10.20729/00204237
2. ClearSy: Atelier B: the industrial tool to efficiently deploy the B method. https://www.atelierb.eu/en/
3. Lecomte, T.: Teaching and training in formalisation with B, FMTea2023 Lübeck, LNCS, vol. 13962, pp. 82–95. Springer, Cham (2023). https://doi.org/10.1007/978-3-031-27534-0
4. Leuschel, M., et al.: The ProB animator and model checker. https://prob.hhu.de/w/index.php?title=The_ProB_Animator_and_Model_Checker
5. Leuschel, M., Butler M.J.: ProB: an automated analysis toolset for the B method. STTT **10**(2), 185–203 (2008). https://doi.org/10.1007/s10009-007-0063-9
6. Schneider, S.: The B-Method: An Introduction. Palgrave Macmillan (2001). ISBN:978-0-333-79284-1
7. Oliveira, M., Lecomte, T.:The B-Method from specification to code, the course of MOOC. https://mooc.imd.ufrn.br/course/the-b-method
8. Wylie, C.R.: 101 Puzzles in Thought & Logic. Dover Publications (1957). ISBN:978-0486203676
9. Onoda H.: Logic puzzle best 100, PHP(2015). ISBN:978-4-569-82545-8
10. Research center for verification and semantics, national institute of advanced industrial science and technology (AIST): model checking -elementary course-, NanoOpt Media (2009). ISBN:978-4-7649-5505-9
11. Ohnishi T., Yamamoto R., Nakamura Y.: A practice of open lecture for elementary and junior high school students to introduce mathematical reasoning. Comput. Softw. JSSST **40**(1), 3–10 (2023). https://doi.org/10.11309/jssst.40.1_3
12. Wikipedia: Sokoban. https://en.wikipedia.org/wiki/Sokoban
13. Hehner, E.C.R.: Computer science - formal methods. https://computingstudy.wordpress.com/formal-methods/
14. Wikipedia: Formal methods in Japanese. https://ja.wikipedia.org/wiki/%E5%BD%A2%E5%BC%8F%E6%89%8B%E6%B3%95
15. Araki K.: New trends in formal methods: formal methods: past, present and future -toward practical applications. Mag. IPSJ **49**(5), 493–498 (2008). https://ipsj.ixsq.nii.ac.jp/records/60933
16. Carvalho, G.: Teaching formal methods for 10 years: reflections on theories, tools, materials, and communities, FMTea2024 Milan, LNCS, vol. 14939, pp. 58–74. Springer, Cham (2024). https://doi.org/10.1007/978-3-031-71379-8
17. Brucker, A.D., Marmsoler, D.: Teaching formal methods in application domains, FMTea2024 Milan, LNCS, vol. 14939, pp. 124–141. Springer (2024). https://doi.org/10.1007/978-3-031-71379-8
18. Aceto, L., Ingólfsdóttir, A.: Introducing formal methods to first-year students in three intensive weeks, FMTea2021 virtual event, LNCS, vol. 13122, pp. 1–17. Springer (2021). https://doi.org/10.1007/978-3-030-91550-6
19. Vanit-Anunchai, S.: Teaching low-code formal methods with coloured petri nets, FMTea2023 Lübeck, LNCS 13962, pp. 96–104. Springer (2023). https://doi.org/10.1007/978-3-031-27534-0
20. Yatapanage, N.: Introducing formal methods to students who hate maths and struggle with programming, FMTea2021 virtual event, LNCS 13122, pp. 133–145. Springer (2021). https://doi.org/10.1007/978-3-030-91550-6

21. Ishikawa F., Taguchi K., Yoshioka N., Honiden S.: What top-level software engineers tackle after learning formal methods: experiences from the top SE project, TFM2009 Berlin, LNCS 5846, pp. 57–71. Springer (2009). https://doi.org/10.1007/978-3-642-04912-5_5
22. Suzuki Y.: National institute of technology, numazu college: 2025's syllabus of "theory of software verification". https://syllabus.kosen-k.go.jp/Pages/PublicSyllabus?school_id=22&department_id=15&subject_code=2025-438&year=2022&lang=ja

Experiential and Practice-Oriented FM Education

Learning Formal Methods Through Project-Based Modeling of Concurrent Systems with Anemone

Manel Barkallah[(✉)] and Jean-Marie Jacquet[(✉)]

Nadi Research Institute, Faculty of Computer Science, University of Namur,
Rue Grandgagnage 21, 5000 Namur, Belgium
{manel.barkallah,jean-marie.jacquet}@unamur.be

Abstract. Formal methods are often perceived by students as abstract and disconnected from real software development, particularly when addressing concurrent systems. This paper reports on an experience of teaching formal methods in a Master-level course through project-based learning, using the Anemone workbench and the Multi-Bach coordination language. Rather than relying on small artificial examples, students apply formal modelling to parts of their own projects involving concurrency and coordination. Animation and trace inspection support the exploration of dynamic behaviours and interleavings, helping students connect formal descriptions to intuitive system behaviour. We describe the course context and representative student projects, and report observations on engagement, recurring modelling difficulties, and learning outcomes. We also discuss limitations of the approach, including tool-learning overhead and scalability concerns. Overall, the experience suggests that tool-supported projects can make formal methods more accessible and meaningful for advanced students while preserving their formal foundations.

Keywords: Formal Methods Education · Project-Based Learning · Coordination Languages · Anemone

1 Introduction

Formal methods [3,23] are often introduced to students through relatively small, self-contained examples, which lead them think of an emphasis put on syntax and proof obligations. While this highlights their mathematical rigor, it can make them appear as yet another formal language to learn, disconnected from practical concerns. As a result, some students struggle to perceive their relevance for real-world software systems [16]. In particular, even when working on systems that are inherently concurrent, reactive, or distributed, students tend to rely primarily on testing, despite its non-exhaustive nature, and may not immediately see how formal methods could help them uncover subtle bugs. Yet concurrency is no longer an advanced or optional topic: it is a fundamental characteristic of

G. Carvalho and T. Kobayashi (Eds.): FMTea 2026, LNCS 16566, pp. 93–110, 2026.
https://doi.org/10.1007/978-3-032-26743-6_6

contemporary software systems, ranging from mobile and web applications to reactive and context-aware systems. Reasoning about interaction, coordination, and interference between concurrent components therefore provides a natural entry point for formal methods education. In this paper, we report on an experience of teaching formal methods through project-based learning in a Master-level course focused on reactive and concurrent systems. Formal modelling is introduced as a means to reason about concurrency and coordination using the Multi-Bach coordination language [4] and the Anemone workbench [14]. Rather than treating formal methods as an isolated topic, they are embedded in student projects that already involve reactive behaviour and interacting components.

This experience takes place in the Master course *Design of Reactive Applications* at the University of Namur (Belgium). The course combines 30 h of lectures with 15 h of project sessions and is delivered to Master students in computer science. It focuses on designing and implementing reactive, context-aware systems and is structured in two parts. The first part covers reactive programming, introducing change propagation, failure handling, and latency, with illustrations in Scala using Functional and Object-Oriented Reactive Programming [6,17]. The second addresses contextual programming and coordination, including the actor model and shared-space coordination; links with declarative frameworks such as React Native are also discussed. Assessment includes a group project in which students model and analyse part of their system using Multi-Bach and Anemone [14]. This project-based approach presents formal methods as practical tools for reasoning about concurrency and coordination rather than abstract theory. Students come from diverse Master programmes (60 and 120 credit tracks, including Data Science and Software Engineering), resulting in a heterogeneous cohort that reflects realistic modelling contexts. To support this approach, in the course subsequently taken over by Corentin Reuther, we rely on coordination languages and executable modelling frameworks that make interaction explicit and concurrency observable.

This paper makes three main contributions. First, it proposes a structured pedagogical formalisation of modelling principles for teaching concurrency. The approach organises modelling activities around a small set of explicit principles that help students move from sequential intuition toward clearer reasoning about concurrent behaviour. Second, it identifies recurring concurrency patterns that emerged across student projects. Making these patterns explicit provides a reusable vocabulary for discussing coordination mechanisms and sources of interference. These patterns were not predefined templates but were progressively distilled from classroom practice. Third, it reports on a three-year experience integrating formal modelling into a course not primarily dedicated to formal methods, highlighting both challenges and pedagogical benefits of embedding lightweight formal reasoning within project-based learning.

Related Work in Teaching FM and Concurrency

FMTea landscape. Recent FMTea papers emphasise reusable teaching experiences and tool-supported learning. FMTea'24 [21] reports the integration of

GitHub Classroom in an FM module, didactic use of contracts in Event-B, educational support for temporal features in Alloy 6, and the use of Logika for constructing correct software [7,19,20,22]. Our work aligns with this perspective and focuses on making concurrency observable in class.

Teaching concurrency via model checking. Model checkers are widely used to expose interleavings and reason about properties. With SPIN/Promela, students simulate executions and explore state spaces; with TLA+, they specify behaviours and check invariants or liveness; with Alloy, they benefit from lightweight relational modelling and automated analysis [12,13,15]. These approaches typically start from properties to be verified. In contrast, we adopt an observation-first approach: students first explore executions and discover non-determinism through traces, and only then reason about properties when needed. This contrasts with more traditional presentations of concurrency, often centred on shared-memory models and synchronisation mechanisms [10], where interleavings are introduced at a lower level of abstraction.

Actor-based teaching. Actor-oriented courses (Akka or Erlang/OTP) help students internalise message passing, distribution, and supervision [2,11]. They provide a clear programming model for concurrent systems, but interactions remain structured around explicit communication between actors. Our stance is complementary: we foreground data-driven coordination over a shared space, allowing students to reason about dependencies, partial observations, and interference without assuming explicit communication paths or a central controller.

Tuple spaces and Linda. Our foundation is close to the Linda family: generative communication and shared tuple spaces as coordination media [9]. Multi-Bach belongs to this lineage; Anemone provides the executable workbench used in our course. Other coordination models, such as Reo [1], similarly explore structured interaction mechanisms between concurrent components. Compared to these approaches, our focus is not on expressiveness but on pedagogical clarity. Our contribution is therefore primarily educational: we distil modelling principles and recurring patterns, and show how animation and trace inspection help students move from sequential intuition to explicit reasoning about interleavings and coordination [14].

2   Anemone as a Teaching Tool

Anemone [14] is used in our course as an executable modelling environment for describing concurrent behaviour through interacting processes and a shared space. Coordination languages were introduced to address a central difficulty of concurrent and distributed systems: how to structure interaction between independent components without entangling it with their internal computation. Rather than relying solely on message passing or explicit control flow, these languages propose a shared coordination medium through which processes interact indirectly.

The seminal *Linda* model [9] introduced the notion of a *tuple space*, a shared associative memory where processes communicate by adding, reading, or withdrawing data using simple primitives such as *out, in,* and *rd.* This abstraction separates computation from interaction and makes synchronization explicit at the level of shared data. Instead of reasoning about who sends a message to whom, designers reason about how the shared space evolves. Building on this idea, *Multi-Bach* [5,14] extends the coordination model with guarded primitives and explicit tests for presence or absence of data. Its core operations - *tell, ask, nask,* and *get* - allow agents to add information to the shared space, check whether certain data is present or absent, and remove it when needed. These primitives make synchronization conditions explicit. They naturally support to be combined with sequential composition (;), parallel composition (∥), and non-deterministic composition (+) operators. This is particularly relevant for modelling reactive and IoT-inspired systems, where multiple components evolve concurrently and interact through a common environment. Other coordination frameworks, such as Klaim [8] and ReSpecT [18], similarly emphasise programmable coordination infrastructures. In all these approaches, the key idea is to provide a structured interaction space that makes concurrency visible and analyzable.

Multi-Bach models systems as collections of concurrent agents that interact through a shared multiset store. The language deliberately separates internal computation from coordination. Agents are defined by behavioural rules, while synchronization occurs exclusively through operations on the shared space. The four core primitives are straightforward, as presented in Fig. 1: *tell(t)* inserts a token t into the store; *ask(t)* checks whether t is present; *nask(t)* checks that t is absent; and *get(t)* removes t from the store. These operations can be combined using standard concurrency operators such as sequential composition, parallel composition, and non-deterministic choice. Structured data allow the representation of complex states. This minimal set of constructs is expressive enough to capture interference, synchronization constraints, and resource dependencies, while remaining sufficiently simple for educational purposes.

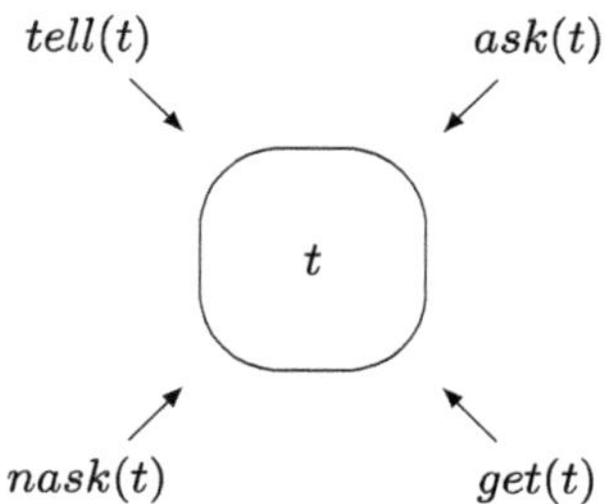

Fig. 1. Primitives of the Multi-Bach Language [14].

Anemone [14] provides an executable environment for Multi-Bach specifications. It combines textual modelling with interactive animation, allowing users to observe how agents interact through the shared space. Models can be executed step by step, with visual feedback showing the evolution of the store and the activation of concurrent actions. This immediate observability plays a central pedagogical role: students can see alternative interleavings, detect unexpected behaviours, and reflect on coordination choices. Beyond execution, Anemone also supports basic verification tasks. Its design deliberately prioritises clarity and accessibility over expressive completeness, making it suitable for teaching contexts where the objective is to understand concurrency mechanisms rather than to perform large-scale verification.

To reason about system evolution over time, Anemone integrates a lightweight fragment of Linear Temporal Logic (LTL) [3]. Propositional formulas describe the presence of structured tokens in the shared store, while temporal operators such as **Next** and **Until** allow users to express progress and safety properties. For example, a reachability property such as

$$Reach(\#finish = 1)$$

expresses that the system eventually reaches a state containing the token *finish*. Note that $Reach(\#finish = 1)$ is equivalent to *true Until* $(\#finish = 1)$. In order to preserve decidability and keep reasoning manageable in an educational setting, verification is performed over a bounded state space. Limits are imposed on the number of tuples and recursion depth. This bounded analysis is sufficient to support discussions about synchronization, interference, and eventual progress, without overwhelming students with the full complexity of unbounded model checking.

From a teaching perspective, Anemone offers several advantages. It provides precise formal semantics while remaining accessible to students who are not specialists in formal methods. Models are expressive enough to capture coordination and interference between concurrent components, yet simple enough to be understood and manipulated by students with a software engineering background. Executing and animating models is central to learning: students observe interleavings, unexpected interactions, and emergent behaviours that are difficult to grasp from static specifications. This immediate feedback helps them connect formal descriptions to intuitive system behaviour. Anemone integrates naturally into a course that is not primarily dedicated to formal methods, presenting modelling as a practical means to structure and reason about complex reactive and concurrent systems rather than as an end in itself, reducing resistance and fostering engagement.

Anemone supports hands-on exploration of concurrent systems. Students can iteratively design, modify, and simulate interacting processes, with visual feedback on how components coordinate and interfere. Trace inspection helps detect subtle concurrency issues such as race conditions and deadlocks, while animations make emergent behaviours tangible and easier to discuss in project teams. The tool lightweight syntax and modular design allow students to extend or adapt models for various contexts, such as multi-agent simulations or reactive applications. Execution traces and animations can be saved, shared, and analysed across groups, promoting peer review and reflection. By bridging formal modelling with practical experimentation, Anemone lets students experience the consequences of design decisions in a controlled environment, making abstract concepts of concurrency and coordination concrete without overwhelming them with complex verification.

3   Student Projects

The projects were carried out in small groups, modelling interactive systems inspired by realistic scenarios, often related to ambient or IoT applications. The goal was to develop coherent formal models capturing concurrent behaviour and coordination. Students defined concurrent processes representing user actions, autonomous activities, and environmental dynamics, all interacting through a shared space. One representative project modelled a connected home centred on a smart mirror aggregating information from various sensors. Other projects followed similar structures, combining interaction, background activities, and multiple views over a shared state. These settings allow students to experiment with coordination and reason about concurrent executions. In the following sections, we present representative projects illustrating how concurrency was modelled and explored using Anemone 2.

Each group delivers: (i) a concise model with comments, (ii) 2–3 short excerpts from tracers or screenshots showing key interleavings, and (iii) a brief note explaining a coordination issue they observed and the policy they adopted to address it. This structure keeps the focus on understanding concurrency rather than on implementation details.

3.1   Modelling Principles

To support systematic reasoning about concurrency, modelling activities are organised around a set of guiding principles, as depicted in Fig. 2. These principles help students transition from intuitive descriptions of interactive systems to structured formal models.

Principle 1: Autonomous Agents. System components are represented as autonomous agents with local behaviours. There is no central controller: global evolution emerges from the concurrent execution of independent agents. This decomposition encourages students to abandon sequential reasoning and analyse interactions between parallel activities.

Principle 2: Shared Coordination Space. Agents do not interact indirectly but by means of a shared coordination space storing structured data. Instead of focusing on control flow, students examine how state evolves and how coordination constraints regulate access and modification. The shared space becomes the central abstraction for understanding interaction.

Principle 3: Data-Driven Interaction. Behaviour is defined by conditions over the shared state. Actions occur when relevant data is available, making dependencies explicit. This perspective highlights how coordination emerges from local reaction rules rather than predefined execution sequences.

Principle 4: Non-deterministic Execution. Concurrent systems allow multiple possible execution orders. Models are therefore analysed under different interleavings, prompting students to reason about properties that must hold across executions rather than for a single scenario.

Principle 5: Observable Interleavings. Models are explored through simulation and trace inspection. Observing state transitions and overlapping actions makes concurrency tangible and reveals unintended interactions. Animation exposes the concrete consequences of modelling choices, supporting formal reasoning.

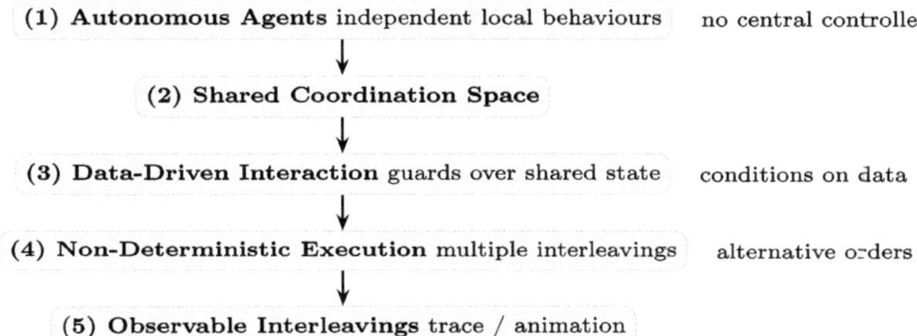

Fig. 2. Summary of the five modelling principles. The numbers and the arrows represent successif reffinements modelling.

3.2   Concurrency Patterns

Beyond the principles above, recurring structural patterns emerged across student projects. Making these patterns explicit proved valuable for two reasons: they offer reusable modelling strategies and provide a shared vocabulary for discussing design choices, coordination mechanisms, and sources of interference.

Reactive Agent Pattern. A reactive agent continuously monitors the shared space and performs actions when certain conditions hold. Its behaviour is defined by guarded rules triggered asynchronously, depending on the evolving state. Students gradually understand that responsiveness can be achieved without central supervision, using well-defined local rules.

SensorActuator Pattern. Sensor agents introduce information into the shared space, while actuator agents interpret that information and produce effects. This clear separation clarifies the flow of information and helps identify potential coordination issues, especially when multiple actuators depend on the same data or updates occur concurrently. The pattern often exposes implicit timing assumptions.

Environmental Agent Pattern. Environmental agents model ongoing background dynamics such as time progression, external events, or stochastic changes. Unlike user-driven components, they evolve independently, encouraging students to see the system as an evolving environment rather than a closed reactive structure. This perspective naturally leads to discussions about interference between background processes and interactive behaviour.

Observer Pattern. Observer agents read the shared space without modifying it. They may record traces, compute derived information, or provide visual representations. This pattern highlights that multiple perspectives on the same evolving state can coexist and raises questions about the timing and stability of observations in a concurrent system.

Orchestration without Sequential Control. Across projects, global behaviour emerges without a dedicated controller. Coordination results from local rules combined with constraints in the shared space. For students used to sequential programming, this is a turning point: correctness depends on consistent coordination policies rather than a prescribed global order (Fig. 3).

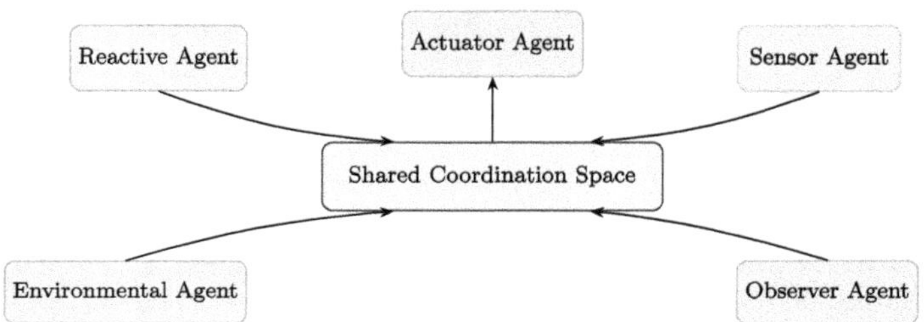

Fig. 3. Concurrency patterns around a shared coordination space. Autonomous agents interact indirectly via the space; global behaviour emerges without a central controller.

4 Formal Methods in Teaching: Experience with Concurrency Reasoning

In this section, we describe teaching formal methods through concurrency reasoning, using project-based learning and the Anemone workbench to explore how interacting components produce complex system behaviours.

4.1 BRACELEDS: Concurrency in a Reactive Concert Environment

The Braceleds project, developed by four students[1], models an interactive concert where devices and participants evolve in parallel. The system relies on a shared coordination space containing tuples such as `music(type)`, `vib(intensity)`, `scent(type)`, and `free_pos(x,y)`. Agents include a conductor (`PlayMusic`), moving participants (`MovePos/Accelerometer`), vibration sensors, floodlights, bracelets, and a scent diffuser. Each component reacts to data in the shared space rather than calling others directly.

---

[1] Yassine Bouncer, Romain Defesche, Carine Pochet, and Kevin Schweitzer.

Concurrency emerges from autonomous participant behaviour and reactive devices operating simultaneously. Each bracelet updates its position nondeterministically, simulating crowd movement. These updates influence the vibration sensor, which writes intensity values into the space. Lighting systems and bracelets react to both **music** and **vib**, while the diffuser monitors musical context and activates **scent(chocolate)** whenever **music(love)** appears. The **PlayMusic** agent structures concert phases but does not impose sequential control. Sound, lights, sensors, and participant movement proceed in parallel, producing non-trivial interleavings, as shown in Fig. 4.

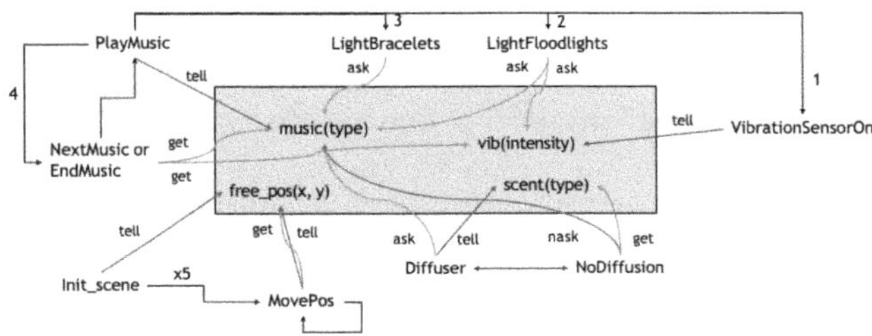

Fig. 4. BRACELEDS project: Agent PlayMusic.

```
PlayMusic =
    SpeakerOn;
    VibrationSensorOn;
    LightFloodlights(floodlight1, floodlight2);
    LightBracelets;
    (NextMusic + NextMusic + NextMusic + EndMusic).
NextMusic =
    (get(music(sad)) + get(music(happy)) + get(music(love)) );
    (get(vib(low)) + get(vib(high)) ) ;
    PlayMusic .
EndMusic =
    SpeakerOff ;   VibrationSensorOff ;   LightsOff .
```

This code defines the behaviour of the **PlayMusic** agent, which structures the progression of a concert while leaving other components to evolve concurrently. The agent first activates the main components of the system: the speaker, the vibration sensor, the floodlights, and the bracelets. It then enters a non-deterministic phase where either a sequence of music transitions (**NextMusic**) or the termination of the concert (**EndMusic**) can occur. The **NextMusic** behaviour models the transition between musical phases. It waits for the presence of a current music state (e.g., **music(sad)**, **music(happy)**, or **music(love)**) and a vibration level (low or high), both retrieved from the shared space. Once these conditions are met, the agent recursively returns to **PlayMusic**, enabling continuous evolution of the system. Finally, **EndMusic** represents the termination phase, where all active components are switched off. Overall, this agent organises the global flow of the concert without enforcing a strict execution order, allowing other agents (lights, sensors, bracelets) to react independently based on the shared state.

Execution traces revealed behaviours that differ from students initial expectations. Lights sometimes reacted before a new `vib` value was written; scent briefly persisted after music changed; bracelet movements altered vibration readings mid-phase. Repeated simulations from identical initial states produced different intermediate configurations. Students initially interpreted these variations as modelling errors. Exploration of traces clarified that they result from non-deterministic scheduling and loose coupling over the shared space.

Simultaneous bracelet updates raised questions about global consistency. At a given instant, the vibration level may reflect only part of the crowd. Rather than inconsistencies, these are partial observations inherent to concurrent systems. The project helped students distinguish logical faults from concurrency artefacts and understand that global properties depend on coordination policies rather than centralised control.

Students progressively adopted simple coordination rules: readthenupdate patterns for lighting, idempotent writes for `music` and `vib`, and a reactive loop (`Diffuser/NoDiffusion`) that watches `music(love)` instead of being called imperatively. They also constrained participant moves to reduce unrealistic churn while preserving non-determinism. The corresponding code is shown below.

```
Diffuser      = ask(music (love));   att(state, diffuser, concert, on);
                tell(scent ( chocolate ) );   NoDiffusion .

NoDiffusion = nask(music (love));   get(scent (chocolate) ) ;
              att (state, diffuser, concert, off); Diffuser.
```

This code implements a reactive control of the diffuser based on the presence of `music(love)`. The `Diffuser` agent activates the diffuser and introduces `scent(chocolate)` when this condition holds, while `NoDiffusion` handles the opposite case by removing the scent and switching the diffuser off. Together, they form a simple loop that continuously reacts to changes in the shared state, without relying on explicit control from other components.

4.2   Smart Mirror: Coordinating a Reactive Home Environment

The Smart Mirror project, carried out by four students[2], models a reactive home environment in which multiple agents representing a user, household sensors, and autonomous processes interact dynamically. The system includes concurrent agents for the user, time progression, dust appearance, and interactive input via a mirror or mobile application. All agents coordinate indirectly through a shared space that represents the current state of the home and its sensors.

Concurrency emerges naturally at several levels. First, the user agent (Saria) evolves autonomously based on input actions, including moving between rooms, toggling devices, and updating a To-Do list or calendar. Second, autonomous agents, such as the time agent and the random dust generator, evolve concurrently, updating the shared space independently of Saria actions. These agents

---

[2] Aline Boulanger, Louise Delpierre Stasser, Margaux Leleu, and Antonin Rousseau.

continuously modify environment state time advances, dust appears or disappears, while the user performs other tasks, leading to non-trivial interleavings. Finally, display agents for the mirror, phone, and clock read the shared space and update visualizations independently, providing feedback that depends on the concurrent activities of all agents. The corresponding Multi-Bach code of the **Saria** agent is shown below, and its structure is illustrated in Fig. 5.

Fig. 5. Smart Home project: left, multiple concurrent agents interacting within a shared environment; right, example of a Smart Mirror integrated into the home setting.

```
Saria  (place:  Place)=
       (  tell(Todo);   show(todo_icon ,mirrorScene);
          show(todo_icon ,phoneScene);
          Saria(place))

    +  (  get(Todo);  Remove_todo_icon;Saria(place))
    +  (  (place=dehors) -> ((Saria(dehors))+ show(saria ,houseScene);
          (Saria(entree))) )
    +  (  (place=entree) -> (Entree) )
    +  (  (place=salon) -> (Salon) )
    +  (  (place=buanderie) -> (Buanderie) )
    +  (  (place=salle_a_manger) -> (Salle_a_manger) )
    +  (  (place=cuisine)-> (Cuisine) )
    +  (  (place=chambre) -> (Chambre)  )
    +  (  (place=salle_de_bain)-> (Salle_de_bain)  ).
```

This code defines the behaviour of the **Saria** agent, which represents a user moving between different locations. The agent reacts to events such as adding or removing a **Todo** item, updating icons accordingly on different interfaces. It also models movement between places (e.g., **dehors**, **entree**, **salon**), where each location triggers a corresponding behaviour. The use of non-deterministic choice allows the agent to alternate between interaction (e.g., managing tasks) and movement across the environment.

This project models a connected home centred on a smart mirror, complemented by a phone and a clock view. The shared space aggregates user actions and home "signals" (e.g., keys, wallet, lights, heater), together with a time tuple **time(hour,day,month,year)** maintained by a dedicated **PassTime** process; a

`Random_vacuum` process injects dust events stochastically. Scenes (house, mirror, phone, clock) are animated independently and read the shared state asynchronously. Non-determinism arises from the coexistence of user-driven steps (`Saria`) and background agents (`PassTime`, `Random_vacuum`). Displays (mirror/phone/clock) are independent observers: they update on their own schedule based on what they read from the shared space. As a result, temporary desynchronisations and partial observations are common, and expected.

Animation made interleavings tangible: a light might toggle just as `PassTime` advances; the mirror shows more detail while the phone keeps a concise subset; the clock recomposes digits by attributes (tens/units) rather than pre-rendering every hour, which surfaces ordering effects between attribute changes and repositioning. These traces clarified that different "views" over the same evolving state will not refresh in lockstep.

Students framed simple policies per view: eventual refresh for the phone (essential indicators only), on-demand or more frequent refresh for the mirror (richer state), and attribute-based updates for the clock to avoid combinatorial widget blow-up. They also used small "generic" tuples (e.g., `Todo`, `Lumiere`) as aggregates to prevent flicker when items appear/disappear in quick succession.

4.3 HYDROPONICS: Coordinating Resources and Human Intervention in a Reactive System

The Hydroponics project, developed by two students[3], models a simplified soil-less cultivation system in which water, nutrients, temperature, and human intervention interact dynamically. The system is structured around a shared coordination space where components publish and react to states such as `water_empty`, `low_nutr`, `hot_stressed`, or `scr_low_water`, as shown in Fig. 6. Core entities include the water reservoir (`Jar`), the plant, a nutrient indicator, a temperature controller, a fan, a LED indicator, a phone interface, and a human actor (`Man`). Each entity is modelled as a finite-state component and evolves by reading from and writing to the shared space. Interaction is entirely data-driven: no component calls another directly.

Fig. 6. Hydroponics Project: Jar fill levels, possible plant conditions and normal state.

---

[3] Grégoire Höllich and Ulrich Touji Nana.

Concurrency arises from the coexistence of environmental dynamics, automatic control logic, and human intervention. Water and nutrient levels evolve through routines (`RoutineWater`, `RoutineNut`), which progressively degrade resource levels. Temperature changes may trigger automatic reactions such as LED colour updates, phone notifications, or fan activation. These processes operate independently. For instance, while water decreases through `LowerWater`, nutrient levels may simultaneously degrade through `LowerNut`. Temperature variations can activate the fan while the plant health is being updated due to low water or nutrient stress. Human intervention (`MoveMan`) may occur at the same time, restoring resources and resetting alerts. Because all components rely on the shared space, updates are not globally synchronised. The observable behaviour depends on the interleaving of state modifications.

Although routines such as `RoutineWater` and `RoutineNut` contain local sequencing through `Delay`, they do not impose a global execution order. Water degradation, nutrient depletion, temperature variations, and human intervention are modelled as independent processes that read and update the shared space concurrently. For instance, while `LowerWater` decreases the reservoir level, `SetTemperature(hot)` may trigger LED updates and fan activation. At the same time, `LowerNut` may modify nutrient levels, affecting plant health before water stress is fully processed. The resulting plant state therefore depends on the interleaving of these updates. Trace inspection showed that identical initial configurations may produce different intermediate states: an alert may briefly appear before another stress condition overrides it; the fan may activate while a water restoration is in progress; plant health may oscillate between stress indicators before stabilising. These effects made concurrency observable and challenged students initial sequential reasoning.

```
(1)   SetTemperature(new_temp : Temperature) =
(2)        ( ask(new_temp) )
(3)     + ( nask(new_temp) ;
(4)         ( (get(cold); get(cold_stressed); get(scr_cold))
(5)          + (get(hot) ; get(hot_stressed) ; get(scr_hot))
(6)          + get(normal));

(7)        tell(new_temp) ;
(8)        ((new_temp = cold -> SetPlantHealthAdd(cold_stressed) ;
(9)          SetPhoneScreenAdd(scr_cold) ; SetLedLight(blue))
(10)        + (new_temp = hot -> SetPlantHealthAdd(hot_stressed) ;
(11)          SetPhoneScreenAdd(scr_hot) ;  SetLedLight(red) ;
(12)          Delay(5)   ; SetFanState(fan_on) ;
(13)          Delay(10) ; SetFanState(fan_off) ; SetTemperature(normal) )
(14)        + (new_temp = normal-> SetLedLight(green);
(15)                            SetPhoneScreenNext; SetPlantHealthNext))).
```

This code defines the behaviour of the **SetTemperature** agent, which:

- (2) check whether the temperature is already present in the shared space.
- (3)–(6) handle the case where the temperature is not present: the agent removes the previous temperature state (cold, hot, or normal) together with its associated effects.
- (7) inserts the new temperature into the shared space.
- (8)–(9) define the behaviour for a cold temperature: the system updates plant health, phone display, and LED colour accordingly.

- (10)–(13) describe the hot case, where additional actions are triggered: the LED turns red and the fan is activated and later deactivated using delays, before returning to a normal temperature.
- (14)–(15) correspond to the normal case, restoring standard system behaviour.

Execution traces revealed non-trivial interactions. The plant health may temporarily reflect water stress before nutrient stress is processed. The phone screen may display an alert that disappears shortly after resource restoration. During temperature peaks, the LED may turn red and the fan activates while the water routine continues decreasing the reservoir level. The introduction of a custom `Delay` function, shown below, made temporal effects visible. Delays structure degradation routines but do not impose strict sequentiality across the whole system. As a result, repeated executions from identical initial states may produce different intermediate configurations, especially when water, nutrients, and temperature evolve concurrently. Students initially interpreted these transient states as inconsistencies. Trace analysis clarified that they result from concurrent updates and partial observations over the shared space.

Ensuring global coherence required careful coordination rules. State modification functions such as `SetJarLevel`, `SetIndNut`, and `SetTemperature` not only update their local component but also trigger dependent updates (plant health, phone screen, LED colour, fan state). Because several alerts may coexist (low water, low nutrients, temperature stress), the phone interface follows a prioritised update policy implemented in `SetPhoneScreenNext`. Similarly, plant health is recomputed according to the current combination of stress indicators rather than through direct control by a central agent. Human intervention is modelled as a state machine. Actions such as adding water or nutrients restore resource levels but must reintegrate into the global coordination logic without breaking ongoing routines. This highlights the absence of a central controller: system behaviour emerges from coordination policies encoded in local rules. The corresponding code is shown below.

As in the previous case studies, concurrency was not artificially introduced. It emerged from modelling independent environmental processes, automatic reactions, and human actions within a shared space. The need to implement a custom `Delay` function exposed limitations of temporal control and reinforced the distinction between local sequencing and global concurrency. Students progressively understood that system behaviour cannot be reasoned about as a linear script: it results from interacting state transitions whose ordering is not predetermined. The Hydroponics project therefore provided a concrete setting to reason about resource management, interference between control loops, and coordination without centralised orchestration.

5   Evaluation and Observations

To assess the pedagogical impact, we conducted a qualitative evaluation over three academic years (2023–2025) with 68 Master students from heterogeneous backgrounds (Software Engineering and Data Science, 60- and 120-credit tracks). Most had prior discrete mathematics and programming experience, but few had worked with formal modelling or verification tools. For many, this was their first hands-on experience with executable formal models.

The evaluation combined instructor observations, analysis of submitted models, informal discussions, and anonymous feedback. The aim was to understand how integrating formal modelling into project-based learning influenced students' reasoning about concurrency, not to perform a controlled experiment. Initially, students tended to reason sequentially, assuming a global execution order and expecting one component to "drive" the system. Non-determinism was underestimated: many groups considered only a single expected scenario and were surprised by alternative interleavings in simulation. Temporary inconsistencies were often seen as errors, and some groups over-constrained the system with artificial sequencing rather than relying on coordination constraints. As projects progressed, reasoning evolved. Through iterative modelling, execution, and trace inspection, students began to discuss interleavings, race conditions, and agent interference explicitly. They shifted from procedural descriptions to analysing how global behaviour emerges from local interaction rules. In final presentations, several groups spontaneously used terms like "shared space constraint", "data dependency", or "non-deterministic execution", showing deeper internalisation of concurrency reasoning. Students highlighted the value of animation and trace exploration. Observing agents evolve independently made concurrency concrete and intuitive. Many had not fully grasped the multiplicity of possible executions before running the models. The shared coordination space also helped them reconsider component interactions. While some noted an initial learning overhead, most found that exploration overcame these difficulties. Over the three years, all groups delivered models meeting the project requirements. Instructors observed clearer agent decomposition, more explicit coordination policies, and fewer sequential assumptions than in previous editions. The evaluation remains qualitative, without pre/post-tests or control groups, relying on observations and student feedback. Future work will explore structured instruments to measure conceptual change in concurrency reasoning more systematically.

Several lessons emerged about teaching concurrency through formal modelling. First, students naturally reason sequentially. Even in reactive, distributed systems, they initially assume implicit central control and predictable execution. Making concurrency explicit requires confronting them with alternative interleavings. Execution and animation are decisive: seeing multiple agents evolve independently has more impact than abstract explanations of non-determinism. Second, non-determinism must be experienced, not just defined. Students often believe they understand concurrency in principle, but only when executions produce different intermediate states do they question their reasoning. Observing unexpected but valid behaviours creates productive cognitive tension, prompting deeper discussions about coordination, consistency, and interference. Third, simplicity of modelling constructs is essential. A lightweight language and a shared coordination space lower the barrier to engagement. Students experiment more freely, and formal methods are seen as practical tools rather than abstract obligations. Fourth, project-based integration reduces resistance. Embedding modelling in a broader reactive systems project makes it a tool to clarify design and anticipate coordination issues, rather than an isolated theoretical exercise. Finally, animation supports, but does not replace, formal reasoning. Visual feedback helps students notice surprising behaviours, but meaningful learning occurs when these are interpreted in terms of model structure and coordination constraints. The combination of formal semantics and observable execution appears key to developing a mature understanding of concurrency.

The progression in students reasoning was also reflected in how they approached modelling decisions and abstraction. In several projects, choosing what to represent in the model was not immediate. Some groups initially included too many details, which made the models harder to analyse, while others simplified too much and missed relevant interactions. These issues often became apparent during execution, when certain behaviours were either difficult to interpret or simply absent. Previously, the course was given slightly after the IoT Master lectures and projects, which limited opportunities for students to apply modelling prior to implementation. After the schedule was changed to run at the same time as the IoT courses, under the supervision of Jean-Marie Jacquet, students were able to specify their models and projects in advance, thereby gaining a clearer understanding of the practical value of lightweight verification models. Through iteration and trace inspection, students progressively adjusted their models, focusing more on coordination aspects and on the role of the shared space. Working on their own project scenarios appeared to help in this process, as students could more easily relate the model behaviour to their initial intentions. Overall, the modelling principles provided a useful reference, although finding an appropriate level of abstraction remained a non-trivial part of the work.

6   Conclusion and Future Work

In conclusion, this experience shows that formal methods, which students often see as abstract or disconnected from real software development, can become much more approachable when integrated into actual projects. Using the Anemone workbench and the Multi-Bach coordination language, students applied formal modeling to parts of their own projects involving concurrency and coordination, instead of just small, artificial examples. Tools for animation and trace inspection helped them explore dynamic behaviors and interleavings, making it easier to connect formal descriptions with how the system actually behaves. While there are challenges, like the time needed to learn the tools, overall, project-based learning with tool support seems to make formal methods more engaging and meaningful for advanced students, without losing their formal rigor. Despite its benefits, the approach also has limitations. Modelling concurrent systems introduces a significant cognitive load, and some students struggled to scale their models as complexity increased. In addition, the learning curve of the tool can be challenging, particularly for students with less interest in modelling. The absence of native time primitives in Anemone required workarounds that added complexity to the models. While pedagogically interesting, these aspects sometimes diverted attention from the core concepts of concurrency and coordination. Finally, this approach fits particularly well in courses where concurrency is already a natural concern. Adapting it to more sequential algorithms focused courses would require careful redesign.

This paper presented an experience of teaching formal methods through project-based modelling of concurrent systems. By embedding formal modelling into realistic student projects and using Anemone to make concurrency observable, we helped students better understand interaction, coordination, and nondeterminism. Future work includes refining the integration of time and autonomy in the models, and exploring how similar approaches could be applied using other coordination frameworks. More broadly, we believe that grounding formal methods education in realistic concurrent scenarios is key to making them both meaningful and effective for advanced students.

Acknowledgment. The authors thank the anonymous reviewers for their insightful comments, which greatly contributed to the improvement of this article. They also thank the University of Namur for its support as well as the Walloon Region for partial support through the Ariac project (convention 210235) and the CyberExcellence project (convention 2110186).

References

1. Arbab, F.: REO: a channel-based coordination model for component composition. Math. Struct. Comput. Sci. **14**(3), 329–366 (2004)
2. Armstrong, J.: Erlang – Software for a Concurrent World. In: Ernst, E. (ed.) ECOOP 2007. LNCS, vol. 4609, pp. 1–1. Springer, Heidelberg (2007). https://doi.org/10.1007/978-3-540-73589-2_1
3. Baier, C., Katoen, J.P.: Principles of Model Checking, MIT press (2008)
4. Barkallah, M.: Socio-technical systems. https://pure.unamur.be/ws/portalfiles/portal/110922546/2025_BarkallahManel_these.pdf
5. Barkallah, M., Jacquet, J.-M.: On the expressiveness and efficiency of guarded lists in Bach. J. Logical Algebraic Methods Programm. 142, 101017 (2025). https://www.sciencedirect.com/science/article/pii/S2352220824000713
6. Blackheath, S.: Functional Reactive Programming, Simon and Schuster (2016)
7. Cansell, D., Méry, D.: The event-B Modelling Method: Concepts and Case Studies. In: Bjørner, D., Henson, M.C. (eds.) Logics of Specification Languages. MTCSAES, pp. 47–152. Springer, Heidelberg (2008). https://doi.org/10.1007/978-3-540-74107-7_3
8. De Nicola, R., Ferrari, G.L., Pugliese, R.: KLAIM: a kernel language for agents interaction and mobility. IEEE Trans. Softw. Eng. **24**(5), 315–330 (2002)
9. Gelernter, D.: Generative communication in Linda. ACM Trans. Programm. Lang. Syst. (TOPLAS) **7**(1), 80–112 (1985)
10. Herlihy, M., Shavit, N., Luchangco, V., Spear, M.: The Art of Multiprocessor Programming, Newnes (2020)

11. Hewitt, C., Bishop, P., Steiger, R.: Session 8 formalisms for artificial intelligence a universal modular actor formalism for artificial intelligence. In: Advance papers of the conference, vol. 3, p. 235. Stanford Research Institute Menlo Park, CA (1973)
12. Holzmann, G.: The SPIN Model Checker: Primer and Reference Manual, Addison-Wesley Professional (2011)
13. Jackson, D.: Software Abstractions: Logic, Language, and Analysis, MIT press (2012)
14. Jacquet, J.M., Barkallah, M.: Anemone: a workbench for the Multi-Bach coordination language. Sci. Comput. Program. 202, 102579 (2021). https://www.sciencedirect.com/science/article/pii/S0167642320301878
15. Lamport, L.: Specifying Systems: The TLA+ Language and Tools for Hardware and Software Engineers. Addison-Wesley (2002)
16. Läufer, K., Mertin, G., Thiruvathukal, G.K.: WIP: an engaging undergraduate intro to model checking in software engineering using TLA. In: 2024 IEEE Frontiers in Education Conference (FIE), pp. 1–5. IEEE (2024)
17. Odersky, M., Spoon, L., Venners, B.: Programming in Scala, Artima Inc (2008)
18. Omicini, A., Ricci, A., Viroli, M.: Time-Aware Coordination in ReSpecT. In: Jacquet, J.-M., Picco, G.P. (eds.) COORDINATION 2005. LNCS, vol. 3454, pp. 268–282. Springer, Heidelberg (2005). https://doi.org/10.1007/11417019_18
19. Padalino, L., Panaccione, F.P., Santambrogio, F., Di Nitto, E., Rossi, M.: An educational module for temporal features in alloy 6. In: Formal Methods Teaching Workshop, pp. 75–90. Springer (2024). https://doi.org/10.1007/978-3-031-71379-8_510.1007/978-3-031-71379-8_5
20. Robby, H., Belt, J., J.: Logika: the Sireum verification framework. In: International Conference on Formal Methods for Industrial Critical Systems, pp. 97–116. Springer (2024). https://doi.org/10.1007/978-3-031-68150-9_6
21. Sekerinski, E., Ribeiro, L.: Formal Methods Teaching: 6th Formal Methods Teaching Workshop, FMTea 2024, Milan, Italy, September 10, 2024. Proceedings. Springer Nature (2024). https://doi.org/10.1007/978-3-031-71379-8
22. Soaibuzzaman, R., J.O.: Introducing Github classroom into a formal methods module. In: Formal Methods Teaching Workshop, pp. 25–42. Springer (2024). https://doi.org/10.1007/978-3-031-71379-8_2
23. Zhumagambetov, R.: Teaching Formal Methods in Academia: A Systematic Literature Review. In: Cerone, A., Roggenbach, M. (eds.) FMFun 2019. CCIS, vol. 1301, pp. 218–226. Springer, Cham (2021). https://doi.org/10.1007/978-3-030-71374-4_12

Teaching Frama-C for Cybersecurity

Julien Signoles$^{(\boxtimes)}$ (ORCID)

Université Paris-Saclay, CEA, List, Palaiseau, France
`julien.signoles@cea.fr`

Abstract. This article presents my feedback on introducing Frama-C for cybersecurity in several engineering and master's level university courses. It consists of a single lecture and a single lab session. Frama-C is an open source platform providing C code analyzers based on formal methods. The course focuses on its three main techniques provided through dedicated analyzers, namely deductive verification with Wp, abstract interpretation with Eva, and runtime annotation verification with E-ACSL. The aim of the course is to introduce students to these formal techniques, emphasizing their practical aspects in cybersecurity.

1 Introduction

Frama-C is an open-source framework for C code analysis [3,23], developed since 2005 and now well established. It can be used both in academia and deployed for industrial purposes. In particular, Frama-C was initially used to verify safety properties of critical embedded applications, for example in avionics [7], nuclear systems [29], space systems [9], or automotive systems [1]. It is now also used to verify security properties. For instance, Thales recently certified a JavaCard Virtual Machine at Common Criteria levels EAL6EAL7 (the highest security certification levels in Europe) using a methodology based on Frama-C and recognized by the French National Cybersecurity Agency (ANSSI) [14–16].

For many years, Frama-C has been used for teaching purposes[1]. In this context, it is generally used during practical lab sessions within courses related to formal methods, such as deductive verification [20], abstract interpretation [12], or runtime annotation checking [21]. Runtime annotation checking is sometimes also included in testing teaching modules. Some uses in cybersecurity curricula have also been reported, for example in the cybersecurity master's program at Université Grenoble Alpes[2]. Nevertheless, experience reports remain rare [13,17,19,33]. In particular, none of them addresses runtime annotation checking or focuses on cybersecurity applications. Also, several tutorial papers have been published [6,22,24,25]. The first focuses on deductive verification, the second on runtime annotation checking, while the third combines both. The

---

[1] A few links to existing courses are mentioned on the very incomplete page https://www.frama-c.com/html/teaching.html.

[2] https://im2ag-moodle.univ-grenoble-alpes.fr/pluginfile.php/69164/course/section/10508/slides_framac.pdf.

G. Carvalho and T. Kobayashi (Eds.): FMTea 2026, LNCS 16566, pp. 111–126, 2026.
https://doi.org/10.1007/978-3-032-26743-6_7

fourth addresses test generation, which is not covered in the course discussed here and is therefore complementary. In summary, no experience report has yet addressed the use of Frama-C in cybersecurity teaching modules, while including the three main verification techniques available in the framework, namely deductive verification, abstract interpretation, and runtime annotation checking.

This paper aims to fill this gap by reporting on the experience gained while teaching an introduction to "Frama-C for Cybersecurity." The course takes the form of a lecture and a lab session, each lasting three hours in their standard versions and one and a half hours in their shortened versions. It was created at the request of Télécom SudParis at the end of 2022. Since then, it is regularly taught in three engineering schools (Télécom SudParis, CentraleSupélec Rennes, and École des Mines Paris) and two master's programs at Université Bretagne Sud ("Cyberus Erasmus Mundus Joint Master in Cybersecurity"[3] and "Cybersecurity of Embedded Systems"[4]), thus training nearly one hundred students per year. It is usually integrated into teaching modules in cybersecurity or embedded systems tracks. The course is given either in English or in French. The companion material (slides and exercises) is in English.

The outline of this paper is as follows. First, Sect. 2 briefly introduces the Frama-C framework. Sections 3 and 4 then present the lecture and the lab session, respectively. Finally, Sect. 5 provides feedback.

2   Frama-C in a Nutshell

Frama-C is an open-source framework for C code analysis. One of its strengths is that it is not a single source code analysis tool, but rather a large collection of analyzers provided as plug-ins, all connected to the platform's kernel and each offering different functionalities. This strength makes it also difficult to grasp and learn, since it is already hard to understand what Frama-C is and provides.

A number of Frama-C's components are shared among all analyzers through the kernel. In particular, user interactions are always performed in a similar way, whether through the graphical interface, named Ivette [10], or via the command line in a terminal. Another essential point is that the analyzed C code can be extended with formal annotations written in the ACSL language, the *lingua franca* of the framework. Finally, the platform allows collaboration between its analyzers, which strengthens its analysis capabilities for advanced use cases.

Figure 1, taken from one of the lecture slides, presents a broad overview of the existing Frama-C plug-ins. Most of them are integrated into the standard Frama-C distribution, while others are distributed externally, and some are closed-source. The figure distinguishes five categories of plug-ins. Verifiers are dedicated to actually checking program properties. They include in particular the three main plug-ins of the platform: Wp [5], based on deductive program

---

[3] https://master-cyberus.eu/.

[4] https://www.univ-ubs.fr/fr/formation-initiale-continue/formations/master-XB/ sciences-technologies-sante-STS/master-cyber-securite-des-systemes-embarques-5SCY00_217.html.

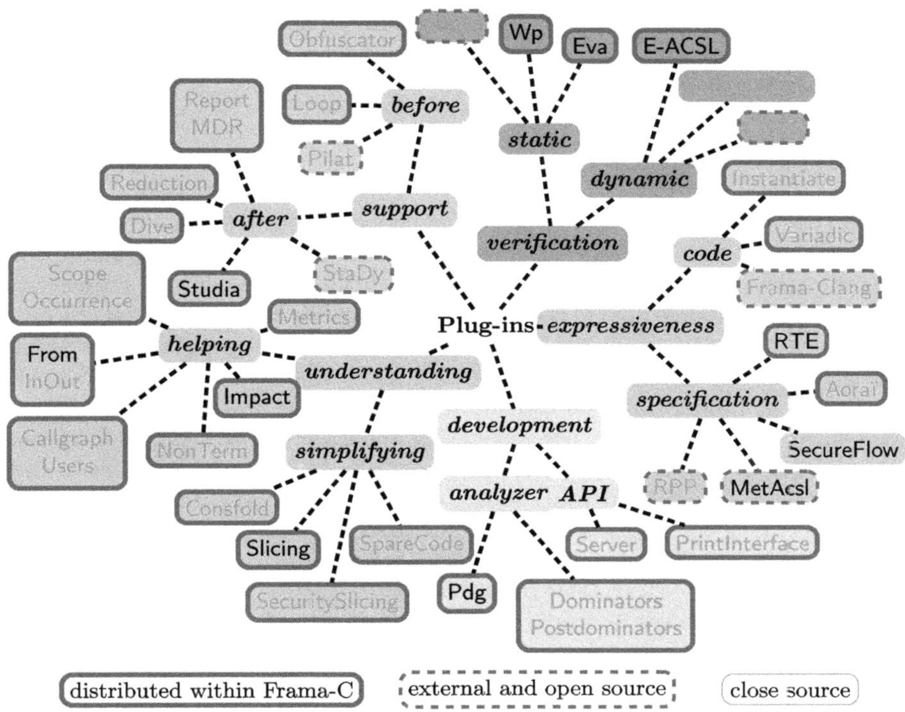

Fig. 1. View of Frama-C analyzers taken from one slide of the lesson: analyzers whose names are written in gray are not explained during the lecture.

verification; Eva [8], based on abstract interpretation; and E-ACSL [4], based on runtime annotation checking. Some plug-ins enhance the expressiveness of analyzed programs or verified properties. For example, regarding analyzed programs, plug-in Frama-Clang supports (a subset of) C++, plug-in Variadic helps other plug-ins analyze variadic functions, and Aorai makes it possible to express temporal properties. For properties, plug-in RTE automatically generates assertions corresponding to undefined behaviors, which correspond to runtime errors in C (e.g., /*@ assert x != 0; */ when the C code divides by x), plug-in SecureFlow focuses on verification of non-interference properties [2], and plug-in MetAcsl is dedicated to verification of properties that are not easily expressible at a single program point, such as confidentiality or integrity properties [32].

Some plug-ins provide services to other plug-ins through their APIs, such as PDG computing a program dependence graph [28]. Several analyzers help users better understand the analyzed code, either by simplifying it (e.g., by slicing it [34] through plug-in Slicing) or by computing specific information, such as dependencies between inputs and outputs with plug-in From or computing an impact analysis with plug-in Impact [27]. Finally, some plug-ins facilitate the use of other analyzers in the platform, either before (for example by obfuscating sensitive industrial code with plug-in Obfuscator) or after an analysis run, for example by highlighting where memory locations are written or read according to Eva in Ivette thanks to plug-in Studia.

3   Lecture Content

The lecture[5] lasts three hours. A shortened version of one and a half hours is also available[6]. This format is not suitable for an in depth learning of Frama-C, nor even of a specific analyzer. Consequently, the chosen approach is to present students with the possible uses of these techniques, while emphasizing that they can be complex to master. The goal is not to train future experts in formal methods. In particular, unlike a full course devoting more than twenty hours to a specific formal technique, such as deductive verification or abstract interpretation, this course does not present the underlying theories. Instead, the goal consists in explaining in concrete terms how formal methods can enhance trust in code security by using practical tools such as Frama-C, but also to highlight their limitations. This way, they will have a few basic keys to decide whether formal method can be a tool of choice to build or enhance trust in software project they might have to deal with in their future career.

The only prerequisite is a minimal knowledge of the C language, sufficient to understand short programs containing pointers and dynamic allocations, without requiring students to write more than a pair of lines of code themselves.

The lecture is divided into six parts. The first part is a preamble that sets the context by justifying the need for tool-supported solutions based on formal methods to increase trust in software. It also informally introduces key qualifiers for code analysis, namely *static, automatic* (and fast), *exact*, and *sound*, while explaining that defining an analysis with all these characteristics is impossible due to Rice's theorem [31]. This allows for a quick introduction to several formal techniques, each of which necessarily relaxes at least one of the above characteristics: runtime annotation checking [21], model checking [18,30], abstract interpretation [12], and deductive verification [20].

After this preamble, the outline is presented: one part dedicated to a brief presentation of Frama-C as a whole; three parts each dedicated to one of the framework's three main verification techniques, i.e., deductive verification with Wp, abstract interpretation with Eva, and runtime annotation checking with E-ACSL, in this order; and a final part devoted to more advanced uses of Frama-C for verifying specific security properties. No part is dedicated to model checking, since this technique is currently little developed within Frama-C. The order of presentation of the three main analyzers is deliberate: techniques are presented from the most complex to the simplest to use. This allows more time to be spent on the first when both students and teacher are still fresh, and to move more quickly through the other two. In particular, the E-ACSL part can be covered very quickly if one run out of time. The first three parts (context,

---

[5] https://julien-signoles.fr/teaching/slides-frama-c-cyber.pdf.
[6] https://julien-signoles.fr/teaching/slides-frama-c-cyber-short.pdf.

Frama-C, and Wp) are covered in the first half of the lecture, and the last three (Eva, E-ACSL, and advanced uses) in the second half.

The second part introduces Frama-C as a whole, emphasizing the platform's choice not to offer a single formal technique, but rather to rely on their diversity and possible combinations. This part notably presents Fig. 1, mainly to explain that none of the grayed-out plug-ins are covered in the lecture or the lab. Beyond the three main plug-ins, the other non-grayed plug-ins are mentioned in the final part of the lecture devoted to advanced uses, except for Studia, which is used in the lab.

The structure of the four technical parts is deliberately always the same: first, an intuition of the underlying technique is provided, without addressing the corresponding theory; then, through examples, usage-specific aspects are presented, emphasizing the main purpose and the main practical difficulties of their uses, as shown on Fig. 2; finally, the technique is illustrated through a demonstration, as shown on Fig. 3, and at least one real industrial use case, as shown on Fig. 4. This choice ensures overall coherence of the lecture while making the techniques more concrete. Moreover, all examples and demonstrations are performed live in front of the students using Frama-C, which makes the lecture more engaging and anticipates aspects that will be revisited in the lab. Student feedback suggests that combining small examples and live demonstrations with real industrial use cases is particularly well received. Another advantage concerns time management, since it is easy to catch up if necessary by skipping live presentation of certain demos or examples. These are included, with their solutions, in the lecture slides, allowing the teacher to present solutions directly if needed and providing students with a self-contained set of course materials.

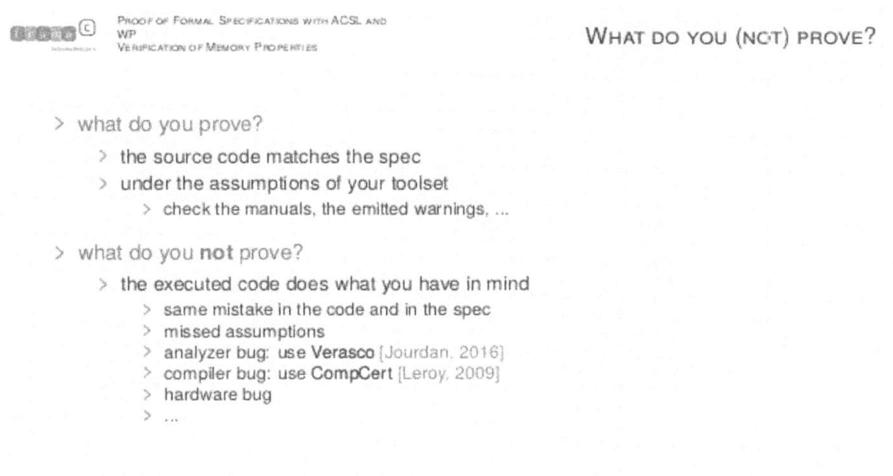

Fig. 2. Slide Highlighting a few Limitations of Deductive Verification.

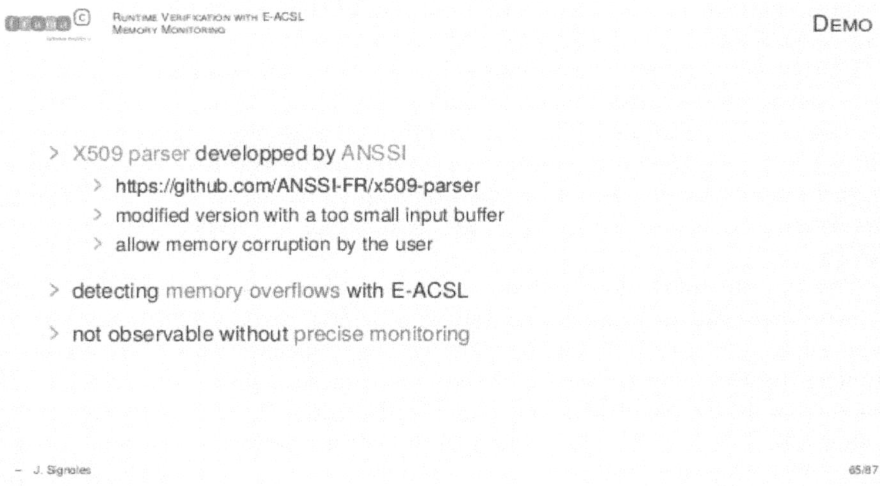

Fig. 3. Slide Introducing a Demo of E-ACSL on a Security-Oriented Software.

The lecture also emphasizes the practical differences between the three techniques, in terms of both advantages and drawbacks. Deductive verification allows complex properties to be verified at the cost of a lack of full automation: users must manually add annotations, and some automatic proofs may fail. Having the teacher perform several examples of increasing difficulty live makes it possible to demonstrate the challenges of program proof without wasting time letting students struggle with problems they are not yet equipped to solve. The examples start with the absolute value function: even if very simple, it is enough to introduce function contracts from pre- and post-conditions to behaviors (case-by-case specification) through frame conditions, as well as proof of absence of undefined behaviors. The second example focuses on memory properties with a function that swaps the content of two pointers. The third example is a variant of this function. It purposely contains a bug. It illustrates how deductive verification deals with function calls, but also the pros and cons of having several memory models relying on different hypotheses, such as absence of aliasing. The final example is a linear search in an array that shows how to deal with loops, including how to write loop variants and loop invariants.

Abstract interpretation is presented more quickly, since students will manipulate the tool themselves during the lab. The lecture shows that this type of tool is well suited to finding undefined behaviors that may correspond to vulnerabilities, while also being able to guarantee their absence when no alarm is raised. However, it emphasizes that even though the technique is automatic, user expertise is required to properly tune parameters in order to achieve a good trade-off between precision, execution time, and memory consumption. Two examples are included, which illustrate the need for trace partitioning [26] and case splitting, respectively. This part also explains how some other analyses based on

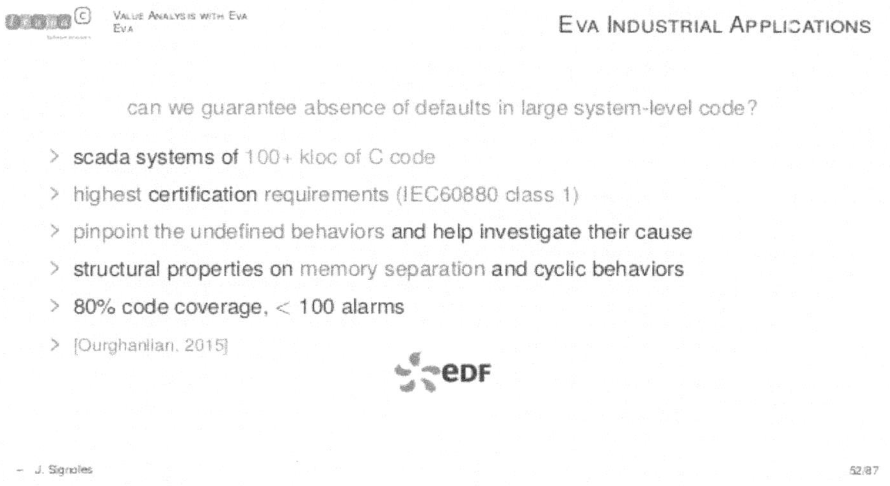

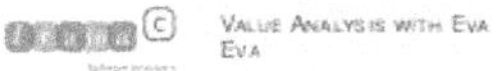

Fig. 4. Slide Introducing an Example of Industrial Application.

abstract interpretation may help security audits, such as slicing, impact analysis and computing dependencies between functional inputs and outputs. Runtime verification of annotations is covered more quickly, as it is fully automatic and does not require particular expertise from the user. Nevertheless, the lecture stresses that a concrete or virtual execution environment is required, and that runtime and memory overheads can be significant in practice.

The final part of the lecture presents several advanced uses in a cybersecurity context. The first one introduces the combination of at least two of the three main verifiers to check, in particular, the absence of undefined behaviors. The second one combines Eva or E-ACSL with the SecureFlow plug-in [2], dedicated to non-interference verification in order to detect that some confidential data leaks on a public channel. It can also detect certain time channel attacks, e.g. when a bit of a secret key is used in a loop condition such that every bit may be eventually read. The last use case combines Wp or E-ACSL with the MetAcsl plug-in [32] to verify integrity and confidentiality properties, typically for operating systems. In each case, the slides only present the target problem, a scheme of the implemented solution without going into any technical detail, and a concrete application: The goal is to present *what* is possible and not *how* it works, which is not possible in such a short format.

The shortened version of the lecture retains the same six-part structure, with each part reduced. Notably, the Wp part contains half less live demos. The essential points remain, but are less frequently reiterated, which may hinder their full assimilation by students.

4   Lab Session

The lab session lasts three hours and is also available in a shortened one-and-a-half-hour version. Its main goal is to explore the capabilities offered by Frama-C, particularly by the Eva plug-in dedicated to value analysis by abstract interpretation. It also aims to give students hands-on experience with the fact that such tools, while powerful, require fine-grained parameter tuning to obtain the desired results.

The students are expected to download and try a provided virtual machine with Frama-C pre-installed before the lab session. They can also install Frama-C by themselves if they prefer. The first version of this lab session was done on the original Frama-C GUI, but it has been moved in 2025 to Ivette [10], the modern Frama-C GUI. The content remains the same, but the explanations have been updated since a large part of the exercises consist in observing Eva's results in the GUI and explaining them.

The lab is organized into two parts: the first dedicated to Eva and the second to Wp. In practice, only the first part constitutes the core of the lab; the second is included solely to provide additional exercises for the very few highly motivated students who choose to work on additional optional exercises at home. Each part consists of two independent exercises: the first one is very easy and helps students get started with the tool, while the second one constitutes the main problem. Each exercise is subdivided into questions that guide students step by step, so they can work alone (almost) without help from the teacher. The shortened version of the lab presents a simplified version of the second Eva exercise and is not detailed here. As with the lecture, the reduced duration leaves less time for students to fully assimilate the concepts.

The first part of the lab is dedicated to Eva. Designing a lab assignment around an abstract interpretation tool like Eva is challenging, because if the analyzed C code is too simple, the analysis is fully automatic and uninteresting, whereas it quickly becomes difficult to tune precisely when the code is even slightly realistic. Finding C code that is both feasible and understandable within a few hours by inexperienced students, while still being interesting, is therefore not easy. The chosen approach is to guide students step by step, in particular by having them observe and explain (with appropriate help in the wording) the results produced by the analyzer. In practice, it is necessary to assist them a bit by answering individual questions during the session, but under these conditions, most students manage to understand the concepts while completing most if not all of the lab within three hours. It is worth noting that solutions are not available online and that LLMs are currently of no help in solving this part of the lab. It is also quite easy to grade because of the step-by-step process that introduces many short questions with a single correct answer possible. If graded, students usually have one week to complete the exercises at home.

The first exercise is based on a small program taken from the NIST's Juliet test suite[7], designed to evaluate code analysis tools. The exercise con-

---

[7] https://samate.nist.gov/SARD/test-suites/112.

sists in using Eva to find the (unique) undefined behavior in the code. Through three questions, it allows students to familiarize themselves with the tool on simple code by observing the results of an unparameterized analysis, and then to improve precision by using the main analyzer parameter, the `-eva-slevel <n>` option, which provides a simple form of trace partitioning [26], one of the main techniques used to improve the precision of abstract analyzers. Once an adequate level of partitioning is specified, the analyzer confirms that it is definitively an undefined behavior. Through this first exercise, students have a first experience with the imprecision of such analyzers when poorly parameterized. Below is the very first question of this exercise, which follows an introduction explaining in particular the possible validity status [11] emitted by Eva:

Question 1
Run Eva in Ivette with the command `ivette -eva cwe126.c`.
Ivette contains several parts. The panel Properties shows the properties corresponding to the ACSL annotations in the code, either manually written or emitted by any Frama-C analyzer (typically, Eva when it raises an alarm). You can click on one of them to show it in the source code.
1. What is the only alarm emitted by Eva? What does it mean?
2. What is its status? What does it mean?
3. Where do the other properties come from? What is their status?

This question is split in several sub-questions that helps the student to observe what is relevant and to conclude. This process implements the step-by-step approach mentioned earlier, which help students to better understand what is going when using the tool.

The second exercise is the main subject of the lab. It consists in analyzing an encryption/decryption module implementing Bacon's cipher from the Rosetta Code project[8], in which I deliberately introduced several errors. Without being correctly parameterized, Eva raises many alarms, which mix true and false alarms. The goal is to use Eva to find the errors, correct each of them by adding or modifying a line of code, and then show that there is no undefined behavior in the program for a large set of inputs, so getting no alarm at all eventually. It is subdivided into five questions.

The first question consists in running the analysis for observing on a particular alarm among those raised. Students re-uses the technique learnt in Exercise 1 to get an unambiguous verdict, then observe the adequate abstract values computed by the analyzer to understand that it is a true alarm and why it happens. They eventually fix the bug to eliminate the alarm.

The second question follows the same pattern for another true alarm, which is inside a loop, so harder to understand. Therefore, they also observe the code using another mechanism provided by Eva, which shows how abstract values evolve when Eva analyzes a loop. It is also an opportunity to observe the fixpoint computation performed by Eva when analyzing a loop. They also use plug-in Studia, which help students understand where the bug comes from in that case.

---

[8] https://rosettacode.org/wiki/Bacon_cipher.

The third question asks students to remove all remaining alarms by improving analysis precision and correcting the code when necessary. It allows students to apply by themselves the methodology underlying the first two questions, while discovering an additional analysis parameter (`-eva-auto-loop-unroll <n>`, which improves precision by unrolling loops n times).

The fourth question is of a different nature, as it asks students to generalize the input context of the analysis using a primitive provided by Eva. Whereas the first three questions actually analyze the program for a fixed string, this question extends the input context to a broad class of strings, typically strings of a fixed given length but containing any alphanumeric character or spaces. The fifth question is similar and further extends the class of considered inputs. It also introduces another analysis parameter (`-ilevel <n>`, which sets the number n of elements up to which sets are represented as discrete sets of n values rather than approximated by intervals covering all values between lower and upper bounds). Starting from a precise input and progressively generalizing the input context in this way is a classical methodology for using Eva, even for operational usage in industrial context.

The two exercises proposed in the second, optional, part on Wp are very classical. The first consists in proving a program that computes a multiplication $x \times y$ as x additions of y starting from 0, while the second aims to verify the implementation of a binary search in a sorted array. The absence of undefined behavior is proven only in the second exercise; the condition guaranteeing the absence of arithmetic overflow in the first exercise is difficult to infer and not particularly interesting. This second part, which is classical and unlikely to be completed by students during the lab, is not detailed further.

5   Feedback

Since I only give a few hours in larger course without being full-time professor in the universities or engineering schools, it is not easy to get a large amount of feedback from students. However, the received feedback of this introductory course on Frama-C for cybersecurity, which is deliberately oriented toward the practical aspects of formal methods, is overall very positive.

5.1   Lesson

During the lesson, the students particularly appreciate the concrete examples presented in the lecture and the industrial applications discussed. The live demo captures their attention and help them stay focus for the other parts as well. They always appreciate being able to fix teacher's slight mistakes (typos, typically) if/when they append.

Their questions are relevant most of the time and allow the discussion of points not covered in the slides, especially regarding C semantics or industrial applications. It is almost always necessary to explain that machine integers are not mathematical integers when talking about integer overflows: it is supposed to have been learnt during their first bachelor year, but it is usually forgotten for many of them. This lesson makes the difference between the two very concrete. Also, they also learn undefined and implementation-defined behaviors, and their possible consequences (in particular, for the former).

Last but not least, I think the most important takeaway of this three-hour lesson consists in highlighting benefits of formal methods, but also their limitations. This way, in their future career, the students will have some clue whether formal methods (and which one) is a potential tool of choice for a particular project. The students seem pretty grateful not to only present the positive side.

5.2   Lab Session

During the lab, the students appreciate solving exercises that are doable without being too easy. They also like working on a real open-source code. During the first years of this course, the exercises were much less subdivide into small questions. As a consequence, they were hardly doable by themselves without teacher's help. It was quite frustrating for many of them, so I have progressively introduced intermediate questions and explanations to help them. They like much more this version than the original one. It is also easier for them to finish the exercises alone at home when necessary. As a consequence, the average grade is about ten-percent higher. During the session, since they work alone more, they also support each other more, asking questions between them about their observation that force them think about what is going on.

However, my advice is still necessary a few times. First, I need to explain the precise abstract representation of pointers in Eva's output, which is explained neither in the lesson nor in the exercises: some of them are fully able to infer it alone, but not all of them. The missing explanation will be added in a further version. Second, the exercises contain an initial important remark (highlighted in a special section) and some hints are also provided when needed. Yet, many students still miss them because they do not always read everything or did not take care enough. Third, Eva's heuristics that improve its efficiency makes it more difficult to understand what the students are observing, e.g., when the number of Eva's loop iterations is slightly smaller than the expected (theoretical) number.

The two last questions (similar in their spirit) are also more complicated since they require they write a few lines of C code by using a Frama-C built-in function in a way they are not familiar with: it takes a bit of imagination, which is not possible for all of them. I think these questions are still necessary. First, they are enjoyable for the best students, who manage to do the other questions quickly. Second, they also show a larger part of the applied methodology when using a value analysis tool such as Eva on concrete code. However, when the lab session is graduated, I often let the last question optional, so as not to penalize students too much if they did not understand that part.

This lab session increase their knowledge of C and secure programming. Typically, they have to understand pointer arithmetic (with the help of Frama-C), to deal with the fact that dynamic allocation functions like `malloc` may return `NULL` and with (simple) error-handling mechanisms.

On the negative side, when working for fixing all remaining alarms, some students "fix" code patterns that are actually not bugs but for which the analyzer generates false alarms when insufficiently parameterized: From their point of view, it is fine since their patches make the considered alarms disappear. Yet, it is usually incorrect since their patches modify the program behavior for statements containing no bug. In such cases, it prevents them to reach 0 alarm, since their incorrect patches prevent to remove alarms at other places in the code but they are not able to do this correlation. I currently have no good way to make it clear for them they are doing something wrong without my support: They usually do not ask for help since everything is fine from their point of view when fixing the alarm they are considering (it is indeed removed).

Another issue is about setting up the lab session: there are sparse but regular issues with installing Frama-C. Indeed, the easiest way to go consists in having Frama-C installed in computers of the lab room but, more on more, the students prefer to use their own laptops anyway. Having the students install Frama-C by themselves may lead to issues, even if provided in a virtual machine. The provided virtual machine does not work on ARM architecture, and so prevent using it on some student's laptop. Sometimes, students have no right to install new software on their laptop (e.g., when they are provided by their engineering school). Also, downloading the virtual machine, whose size is 3+ Gb, takes time: sometimes, the download stops before fully finished and so the students complain the virtual machine does not work. Some students prefer to install Frama-C by their own, which may lead to different problems, e.g., regarding the required dependencies for Ivette. The best way for managing this issue consists in grouping the students pairwise, so it is not a problem if a few laptops are not able to run Frama-C or the virtual machine. However, it shows that democratizing formal methods requires an easy setting, which is not yet fully the case for Frama-C.

5.3   Short Version

The short version of the lesson and the lab session (1.5 h each instead of 3 h each) is not as mature as the other one since it has been created more recently and is taught less frequently.

Shortening the lesson was not very complicated since I mainly removes live demos. I have only kept one or two demos in every part of the course to still have it attractive. The outline and the global message remain the same. However, this message is harder to deliver since I lack time to repeat the key insights regularly during the lesson (and learning requires repetition).

Shortening the lab session requires more work. Basically, it removes the introductory exercise and moves the relevant introductory questions on the main one. The repetitive questions have also been removed to shorten the exercises. It works fine, yet it also lacks the necessary repetition of for good learning.

In its current version, this short lab session is even a bit too easy. It will be extended to add an extra question for the best students, e.g., the first of the two more complicated questions mentioned in Sect. 5.2.

Since the lab session is shorter, it is critical to solve very quickly the installation issues of Frama-C or the provided virtual machine: they cannot afford spending half an hour on installation problem in such a short session

6   Conclusion

This paper has presented how a three-hour lecture about Frama-C for cybersecurity and its associated three-hour lab session is enough to introduce several formal-method techniques and their applications for cybersecurity to students in a practical way, which make them interested. It has also presented their shorter versions (limited to one hour and a half each). Both the lecture and the lab session are now quite mature and only slightly modified, e.g., to keep them up-to-date with Frama-C's new developments. The lab session has been harder to fine tune than the lecture, in particular to obtain exercises that are doable and understandable by students in the given amount of time.

Recently, a colleague of mine used my material (slides and exercises) with success in another university: her feedback was also very positive. This proves that the material is sufficiently self-content to be used by other teachers with good knowledge in Frama-C. Since one year, I have also experimented to give the lecture in different formats and alone, i.e., without the lab session: For instance, I gave it as a one-hour seminar for master students at Université Aix-Marseille, or as a two-hour lecture at the CyberHOT summer school[9]. It has been appreciated by the students as well, even though the abstract interpretation part with Eva remains at a more abstract level because of the absence of lab session: I would recommend to keep this session when possible. However, it has demonstrated that the lecture can easily be adapted to any length from one-hour to three-hour lesson, while preserving its main goal, which is to give them a few basic keys to decide whether formal method can help enhance trust in software project they might have to deal with in their future career.

Acknowledgment. I would like to thank Virgile Prevosto, who originated the initial idea for the lab, and Nikolai Kosmatov, author of some of the lecture's examples. Moreover, this paper would not have been possible without the support of the program coordinators who gave me the opportunity to teach this course in their institutions, in particular Olivier Levillain at Télécom SudParis; Jean-François Lalande and Pierre Wilke at CentraleSupélec Rennes; Salah Sadou and Guy Gogniat at Université Bretagne Sud; and Olivier Hermant at École des Mines Paris. I would also like to thank the anonymous reviewers who provided insightful comments to improve the quality of this paper.

This work was supported by a national grant from the French National Research Agency (ANR) under the France 2030 program, grant number ANR-23-CMAS-0014. The views expressed in this article do not necessarily reflect the views of the French government.

---

[9] https://www.cyberhot.eu/.

References

1. Amilon, J., Gurov, D., Lidström, C., Nyberg, M., Ung, G., Wingbrant, O.: Autodeduct: a tool for automated deductive verification of C code. CoRR abs/2501.10889 (2025). https://doi.org/10.48550/ARXIV.2501.10889
2. Barany, G., Signoles, J.: Hybrid information flow analysis for real-world C code. In: Gabmeyer, S., Johnsen, E.B. (eds.) TAP 2017. LNCS, vol. 10375, pp. 23–40. Springer, Cham (2017). https://doi.org/10.1007/978-3-319-61467-0_2
3. Baudin, P., et al.: The dogged pursuit of bug-free C programs: the Frama-C software analysis platform. Commun. ACM (2021). https://doi.org/10.1145/3470569
4. Benjamin, T., Signoles, J.: Runtime annotation checking with Frama-C: the E-ACSL plug-in. In: [23] (2024). https://doi.org/10.1007/978-3-031-55608-1_5
5. Blanchard, A., Bobot, F., Baudin, P., Correnson, L.: Formally verifying that a program does what it should: the wp plug-in. In: [23] (2024). https://doi.org/10.1007/978-3-031-55608-1_4
6. Blanchard, A., Kosmatov, N., Loulergue, F.: A lesson on verification of IoT software with Frama-C. In: International Conference on High Performance Computing & Simulation (HPCS) (2018). https://doi.org/10.1109/HPCS.2018.00018
7. Brahmi, A., et al.: Industrial use of a safe and efficient formal method based software engineering process in avionics. In: European Conference on Embedded Real Time Software and Systems (ERTS) (2020). https://www.erts2020.org/uploads/presentation-pdf/th1-b3-souyris.pdf
8. Bühler, D., Maroneze, A., Perrelle, V.: Abstract interpretation with the Eva plug-in. In: [23] (2024). https://doi.org/10.1007/978-3-031-55608-1_3
9. Busquim e Silva, R.A., Arai, N.N., Burgareli, L.A., Parente de Oliveira, J.M., Sousa Pinto, J.: Exploring Frama-C resources by verifying space software. In: [23] (2024). https://doi.org/10.1007/978-3-031-55608-1_14
10. Correnson, L.: Ivette: a modern GUI for Frama-C. In: Workshop on Formal Integrated Development Environment (2022). https://doi.org/10.1007/978-3-031-26236-4_10
11. Correnson, L., Signoles, J.: Combining analyses for C program verification. In: Stoelinga, M., Pinger, R. (eds.) FMICS 2012. LNCS, vol. 7437, pp. 108–130. Springer, Heidelberg (2012). https://doi.org/10.1007/978-3-642-32469-7_8
12. Cousot, P.: Principles of Abstract Interpretation. MIT Press (2022)
13. Creuse, L., Dross, C., Garion, C., Hugues, J., Huguet, J.: Teaching deductive verification through Frama-C and SPARK for non computer scientists. In: Dongol, B., Petre, L., Smith, G. (eds.) FMTea 2019. LNCS, vol. 11758, pp. 23–36. Springer, Cham (2019). https://doi.org/10.1007/978-3-030-32441-4_2
14. Djoudi, A., Hána, M., Kosmatov, N.: Formal verification of a JavaCard virtual machine with Frama-C. In: Huisman, M., Păsăreanu, C., Zhan, N. (eds.) FM 2021. LNCS, vol. 13047, pp. 427–444. Springer, Cham (2021). https://doi.org/10.1007/978-3-030-90870-6_23
15. Djoudi, A., Hána, M., Kosmatov, N.: Proof of security properties: application to JavaCard virtual machine. In: [23] (2024). https://doi.org/10.1007/978-3-031-55608-1_16
16. Djoudi, A., et al.: A bottom-up formal verification approach for common criteria certification: application to JavaCard virtual machine. In: European Congress on Embedded Real-Time Systems (ERTS) (2022). https://hal.science/hal-03695829
17. Dubois, C., Prevosto, V., Burel, G.: Teaching formal methods to future engineers. In: Dongol, B., Petre, L., Smith, G. (eds.) FMTea 2019. LNCS, vol. 11758, pp. 69–80. Springer, Cham (2019). https://doi.org/10.1007/978-3-030-32441-4_5

18. Emerson, E.A., Clarke, E.M.: Characterizing correctness properties of parallel programs using fixpoints. In: de Bakker, J., van Leeuwen, J. (eds.) ICALP 1980. LNCS, vol. 85, pp. 169–181. Springer, Heidelberg (1980). https://doi.org/10.1007/3-540-10003-2_69
19. Güdemann, M.: Online teaching of verification of C programs in applied computer science. In: Ferreira, J.F., Mendes, A., Menghi, C. (eds.) FMTea 2021. LNCS, vol. 13122, pp. 18–34. Springer, Cham (2021). https://doi.org/10.1007/978-3-030-91550-6_2
20. Hähnle, R., Huisman, M.: Deductive software verification: from pen-and-paper proofs to industrial tools. In: Steffen, B., Woeginger, G. (eds.) Computing and Software Science. LNCS, vol. 10000, pp. 345–373. Springer, Cham (2019). https://doi.org/10.1007/978-3-319-91908-9_18
21. Huisman, M., Wijs, A.: Runtime annotation checking. In: Concise Guide to Software Verification. Texts in Computer Science. Springer (2023). https://doi.org/10.1007/978-3-031-30167-4
22. Kosmatov, N., Prevosto, V., Signoles, J.: A lesson on proof of programs with Frama-C. Invited tutorial paper. In: International Conference on Tests and Proofs (TAP). https://doi.org/10.1007/978-3-642-3
23. Kosmatov, N., Prevosto, V., Signoles, J.: Guide to Software Verification with Frama-C: Core Components, Usages and Applications. Springer, Cham (2024). https://link.springer.com/book/10.1007/978-3-031-55608-1
24. Kosmatov, N., Signoles, J.: A lesson on runtime assertion checking with Frama-C. In: Legay, A., Bensalem, S. (eds.) RV 2013. LNCS, vol. 8174, pp. 386–399. Springer, Heidelberg (2013). https://doi.org/10.1007/978-3-642-40787-1_29
25. Kosmatov, N., Williams, N., Botella, B., Roger, M., Chebaro, O.: A lesson on structural testing with. In: Brucker, A.D., Julliand, J. (eds.) TAP 2012. LNCS, vol. 7305, pp. 169–175. Springer, Heidelberg (2012). https://doi.org/10.1007/978-3-642-30473-6_15
26. Mauborgne, L., Rival, X.: Trace partitioning in abstract interpretation based static analyzers. In: Sagiv, M. (ed.) ESOP 2005. LNCS, vol. 3444, pp. 5–20. Springer, Heidelberg (2005). https://doi.org/10.1007/978-3-540-31987-0_2
27. Monate, B., Signoles, J.: Slicing for security of code. In: Lipp, P., Sadeghi, A.-R., Koch, K.-M. (eds.) Trust 2008. LNCS, vol. 4968, pp. 133–142. Springer, Heidelberg (2008). https://doi.org/10.1007/978-3-540-68979-9_10
28. Ottenstein, K.J., Ottenstein, L.M.: The program dependence graph in a software development environment. In: Software Engineering Symposium on Practical Software Development Environments (1984). https://doi.org/10.1145/800020.808263
29. Ourghanlian, A.: Evaluation of static analysis tools used to assess software important to nuclear power plant safety. Nucl. Eng. Technol. (2015). https://doi.org/10.1016/j.net.2014.12.009
30. Queille, J.P., Sifakis, J.: Specification and verification of concurrent systems in CESAR. In: Dezani-Ciancaglini, M., Montanari, U. (eds.) Programming 1982. LNCS, vol. 137, pp. 337–351. Springer, Heidelberg (1982). https://doi.org/10.1007/3-540-11494-7_22
31. Rice, H.G.: Classes of recursively enumerable sets and their decision problems. Trans. Am. Math. Soc. (1953). https://doi.org/10.2307/1990888
32. Robles, V., Kosmatov, N., Prevosto, V., Rilling, L., Le Gall, P.: MetAcsl: specification and verification of high-level properties. In: Vojnar, T., Zhang, L. (eds.) TACAS 2019. LNCS, vol. 11427, pp. 358–364. Springer, Cham (2019). https://doi.org/10.1007/978-3-030-17462-0_22

33. Souaf, S., Loulergue, F.: Experience report: teaching code analysis and verification using Frama-C. In: International Workshop on Applicable Formal Methods (appFM) (2021). https://inria.hal.science/hal-03338928
34. Weiser, M.: Program slicing. In: International Conference on Software Engineering (1981). https://dl.acm.org/doi/10.5555/800078.802557

VeHa: A Hybrid National Verification Hackathon for Better Formal Methods Education

Sergey Staroletov[1,2]($\boxtimes$)(iD), Dmitry Kondratyev[3](iD), Vladimir Shelekhov[3,4](iD), Alexander Kogtenkov[5](iD), Nikolay V. Shilov[6](iD), Natalia Garanina[3](iD), Irina Shoshmina[7](iD), Timofey Cherganov[8](iD), and Vasil Dyadov[5](iD)

[1] Institute of Automation and Electrometry, Novosibirsk, Russia
serg_soft@mail.ru
[2] Polzunov Altai State Technical University, Barnaul, Russia
[3] A.P. Ershov Institute of Informatics Systems, Novosibirsk, Russia
[4] Novosibirsk State University, Novosibirsk, Russia
[5] Kaspersky Lab, Moscow, Russia
kwaxer@mail.ru
[6] Lyceum 22 "Hope of Siberia", Novosibirsk, Russia
[7] Peter the Great St. Petersburg Polytechnic University, St. Petersburg, Russia
[8] RusBITech-Astra LLC, Moscow, Russia

Abstract. Developing highly reliable software requires the use of formal verification methods, but integrating them into the education of IT students is difficult due to the gap between theoretical understanding and practical skills. In this paper, we present the design, implementation, and analysis of VeHa-2025, a national verification hackathon aimed at overcoming these challenges through hands-on learning, particularly through competitions focused on specific verification problems. Our event combines tracks in model checking (SPIN, TLA+) and deductive verification (Isabelle/HOL, Coq-Rocq, Why3, C-lightVer). Launched in 2023 as a volunteer initiative by a group of academics, VeHa evolved into an academy-industry partnership by 2025. VeHa-2025 included a structured educational cycle consisting of preparatory workshops, an invited lecture, a three-day competitive problem-solving period, and a post-event verification of solutions. The analysis of metrics, submitted solutions, and participant feedback presented in this article demonstrates the growing effectiveness of our hackathon in developing practical skills, building a community, and equipping participants with competencies in formal methods that are in demand in industry. We argue that such competitive, practice-oriented formats are a valuable pedagogical tool for motivating and training the next generation of software verification engineers.

Keywords: software verification · hackathon · deductive verification · model checking · teaching formal methods

1   Introduction

Formal software verification methods are a fundamental tool for ensuring the reliability of critical systems. However, despite their growing importance, there is a significant gap between the theoretical study of these methods in academia and the practical skills of applying them to real-world problems. As we can state, fundamental disciplines are often less popular among students due to the high entry barrier, the difficulty of choosing first practical projects, and the lack of applied project activities in curricula. At the same time, there is growing interest from the student's side in things related to the semantics of programming languages and software reliability. At universities, formal methods are either combined with software testing courses [44] or taught as specialized or elective courses; this approach is insufficient for the large programming community. To compensate for this imbalance and stimulate practical mastery of formal methods, we, a group of Russian researchers, proposed a hackathon-style competition VeHa (VErification HAckathon, in Russian, "veha" also means "a milestone").

Modern hackathons have established themselves as an effective format for intensive teamwork aimed at solving applied problems in a limited time, traditionally associated with software development in imperative languages. In this paper, we propose to consider the hackathon as an effective way to practically master formal verification methods as a supplement to traditional academic education. In fact, verification using tools like SPIN [24], TLA+ [32,46], or Isabelle/HOL [40] is also a form of *programming*, but in a kind of domain-specific language, where the "code" is divided into *specification* and *proof*, and so can be considered a logical extension of coding tasks. Therefore, this hackathon focuses not on creating a working prototype, but on formally proving the correctness of models and algorithms. The combination of competition, tight deadlines, training materials, and mentoring creates a motivating environment for overcoming the high entry barrier to formal specification and verification. With these goals in mind, we developed the aforementioned series of national (all-Russian) VeHa competitions. The first two competitions were held in the city of Innopolis as a satellite event of the former *Program Semantics, Specification, and Verification workshop* (PSSV), and the most recent competition, described in this article, was held in Novosibirsk in 2025 as a satellite event of the Pottosin Open Siberian Programming Contest [15,23].

In 2023, our first formal verification competition was held in a hybrid (in-person and online) format. Participants, students from Russian universities, were asked to solve problems simultaneously in two main areas: deductive verification using the C-lightVer system [30] for the C-light language (developed by the A.P. Ershov Institute of Informatic Systems) and model checking using the SPIN tool and the Promela language [37]. Verification tasks ranged from relatively simple (for example, deductive verification of array element comparison function) to complex ones based on real-world incidents (such as the modeling of the Russian Luna-25 space station accident, which served as the leitmotif of the first hackathon; the deductive part was also related to it), and industrial protocols. Of the 15 teams, 13 presented solutions for deductive verification with 9 of them receiving maximum scores. However, in the model checking section, most participants encountered significant challenges, such as formulating LTL properties, modeling concurrent interactions, and analyzing counterexamples. Only one team managed to properly solve the Luna-25 model checking task. Feedback from participants at the time indicated overall satisfaction with the event, but also revealed issues with the problem formulations and the need for more mentoring [45].

At the second VeHa-2024 competition, participants were presented with five formal verification tasks: verification of access rights functions in Frama-C [11] and Coq [12, 39]; modeling of a GPU computational pipeline with optimal parameter search through model checking; deductive program verification for the 2-SAT problem using the C-lightVer tool (exercised at the previous hackathon); and construction and verification of an IBFT consensus protocol model [35, 47]. Compared to the first event, there was some increase in registrations and the practical focus of the tasks, but not all tasks were solved – for example, the jury for the consensus problem received no submissions to grade. A total of 16 winners and laureates were announced across various categories, and feedback highlighted challenges with the organization of solution submissions and increased interest in industrial applications of formal methods [31].

In this paper, we describe a further development of VeHa educational initiative, present the results of the third all-Russian VeHa-2025 competition, and discuss lessons learnt from three years of experience. The event was organized by a consortium of academic researchers (from A.P. Ershov Institute of Informatics Systems, Institute of Automation and Electrometry of the Siberian Branch of the Russian Academy of Sciences, Novosibirsk State University, Peter the Great St. Petersburg Polytechnic University) and leading industrial partners (Kaspersky Lab, Astra Group). The collaboration resulted in a proper balance between educational tasks from academia and industrial tasks from the companies that recruit personnel among fresh graduates for departments responsible for software reliability.

The remainder of this paper is structured as follows. Section 2 reviews verification competitions and briefly compares VeHa with a popular International Collegiate Programming Contest. Section 3 details the pedagogical design and structure of the VeHa-2025 event. In Sect. 4, we describe the six competition tasks and explain their educational objectives. Section 5 presents the results of the competition, including an analysis of submitted solutions, some analytics about participants, and feedback. Finally, Sect. 6 concludes the paper with a discussion of lessons learned, limitations, and directions for future iterations of the VeHa series.

When writing the text, the DeepSeek LLM [34] was used to perform the necessary statistical generalizations with mandatory human verification and editing.

2 Related Work

Modern formal methods pedagogy actively utilizes practice-oriented formats, among which competitions and specialized seminars occupy a special place. These events serve as a bridge between theoretical learning and solving engineering problems, consistent with the general trend of challenge-based learning in higher engineering education. Deductive verification competitions have emerged as a distinct educational genre. The VerifyThis series, held since 2011, was initially positioned as a way to engage researchers and practitioners in a discussion of the complexities of software specification, implementation, and verification. The key pedagogical idea is not to evaluate speed, but rather the ability to solve complex software verification problems (Xavier et al., 2025) [18], shifting the focus from competition to depth of understanding. This idea has evolved into long-term challenges, allowing participants to work on realistic industrial problems outside the constraints of conference schedules (Ahrendt et al., 2025) [8]. This reflects the principle of deep learning, which is essential for mastering complex formal concepts. Model checking competitions, such as the RERS series, have also evolved in an educational direction. Initially, the focus was on comparing the performance and capabilities of verification tools (Geske et al., 2016) [21]. However, the emphasis has gradually shifted to creating scalable and controllable synthetic benchmarks (Howar et al., 2021) [25] that can serve as educational material. The introduction of problems from industrial partners (Jasper et al., 2019) [27] has strengthened the connection between training and real-world problems, implementing the principle of authentic learning. The merger of VerifyThis, RERS, and several other well-known contests into a single event has created the TOOLympics competition series [13]. Two events in this series have currently taken place: TOOLympics 2019 and TOOLympics 2023. These are the closest contests to the VeHa series. However, unlike VeHa, TOOLympics events are not annual competitions. SpecifyThis offers a unique educational format. Designed as a workshop to bridge the gaps between different program specification paradigms (Ahrendt et al., 2022) [9], this event directly addresses one of the key pedagogical challenges: the difficulty of formalizing informal requirements. Its case-based format is consistent with case-based learning. Integrating hackathon practices into curricula is recognized as an important area of focus. Research shows that students who participate in competitions demonstrate a deeper understanding of the practical applicability of the methods they learn (Khazeev et al., 2019) [28]. Therefore, we believe the design of our competitions is justified and in line with global trends in formal methods.

As educators, we must compare VeHa with another programming competition among tens of thousands of students worldwide—the International Collegiate Programming Contest (ICPC, until 2017—ACM ICPC [1, 14]), still popular in Russia. The first ACM ICPC finals were held as part of the annual ACM computer science conference, the *5th ACM Computer Science Conference* (CSC '77). Similarly, the first two VeHa competitions, VeHa-2023 and VeHa-2024, were affiliated with the workshop *Program Semantics, Specification and Verification: Theory and Applications* [5,6] partly with a purpose to promote theory of program semantics, specification and verification among VeHa participants. Each ICPC round is organized as follows: a team is assigned one computer for five hours, and all teams are given a single set of 8–12 problems (the problems are written in English in free form). Solutions must be submitted to a testing server to be verified against a large number of different input tests prepared by the jury (but unknown to the participants). A solution is accepted as correct if it passes all tests with the correct result within the time specified for each test, using the amount of memory specified for each test. We believe that neither testing as a verification method, nor the design of expert-based test suites for verifying the correctness of solutions, nor comparison with unverified "jury" solutions, can be considered modern comprehensive methods for assessing the submitted solutions and the qualifications of competition participants. We believe that integrating formal verification into student programming competitions (including ICPC and *Codeforces* [4]) is a pressing need to improve the quality of judging and to enhance the educational level of participants.

3   Educational Framework and Hackathon Design

We decided that the pedagogical foundation of the VeHa competition should be a learning-by-doing approach, enhanced by elements of teamwork and direct interaction between participants, mentors, and industry representatives. A key aspect of the event's philosophy is to create a direct bridge between academic research and industrial needs. This was achieved by engaging employees of leading IT companies (Kaspersky Lab, Astra Group) as organizers and mentors. Following the first hackathon, these employees themselves joined our team in formulating problems based on real-world cases. In the last two hackathons, they ensured the relevance of the technologies being studied to modern challenges in the field of reliable software development, specifically, secure operating systems.

The central educational innovation of the VeHa series is the combination of two fundamental formal verification paradigms – model checking and deductive verification – in a single competition. This hybrid approach allows participants to gain a comprehensive understanding of the range of existing methods. The model checking track (using SPIN/Promela and TLA+ tools) emphasizes the formal description of the behavior of parallel and distributed systems and the automatic verification of temporal properties. In the deductive verification track (using Isabelle/HOL, Coq, Why3, and C-lightVer systems), participants master techniques for constructing formal proofs of the correctness of algorithms relative to their functional specifications.

The VeHa-2025 event was structured as a sequential educational cycle that started on November 5 and concluded with the main competition period on November 8. The preparatory phase included the publication of a full set of educational materials on a dedicated website [7] (developed on the Google Sites platform, based on experience from previous years), see Fig. 1: from mini-manuals and video tutorials for each tool, posted in advance, to well-developed descriptions of the six competition problems, which became available at the start of the

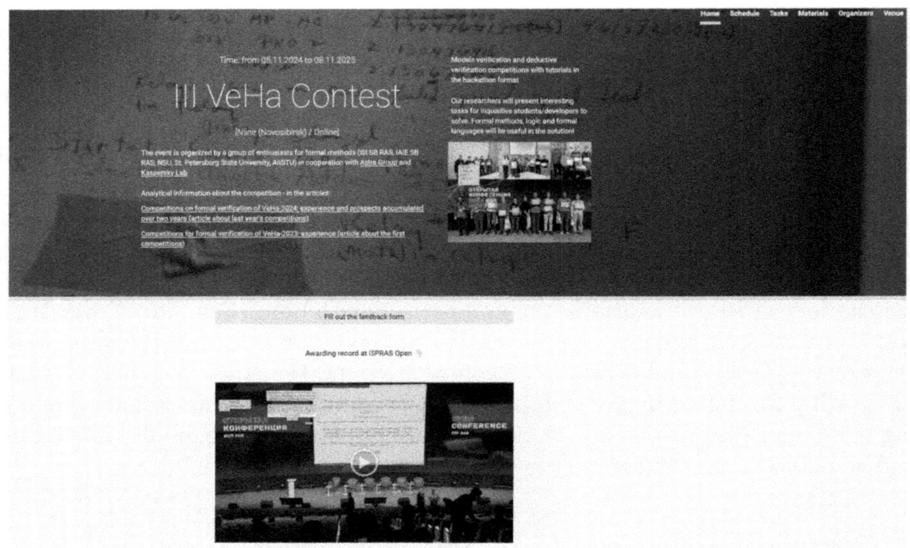

Fig. 1. The event's website [7] (autotranslated into English).

competition. An important component was tutorials or open lectures, conducted in a hybrid format using the domestic Yandex Telemost platform [3].

Let us introduce the content of the tutorials. First, a lecture about the C-lightVer system was given, including a demonstration of deductive verification using the C-lightVer tool. It is important that the participants were able to get acquainted with the organizers' own developments. Next, the hackathon featured an online lecture on the TLA+ tool for formal verification of reactive and distributed systems. The session covered key concepts: TLA+ as a temporal action logic, its advantages for modeling systems with large state spaces (such as network protocols), and a comparison of model checking methods with deductive verification. The practical portion included an interactive demonstration of specification writing—from defining state variables and transition predicates to formulating invariants and temporal properties (the *always* and *eventually* operators). Special attention was paid to pedagogical techniques, such as learning through interactive experimentation in the TLC environment, counterexample analysis, visualization of execution traces, and explanation of abstractions through nondeterminism. An approach to typing through invariants was also demonstrated, and issues related to fairness were discussed. The tutorial combined theoretical explanation with practical examples, allowing participants to quickly grasp the basics of TLA+ and apply them to subsequent hackathon challenges.

Another online tutorial was also held on mastering the Rocq tool (a rebrand of Coq), built on the principle of gradually increasing the complexity of concepts and active learning-by-doing. The session began with an introduction to the basic elements of the language: defining inductive types (such as Booleans and natural numbers) and writing simple functions using pattern matching, allowing participants to immediately apply theory in practice through the interactive *Check* and *Compute* commands. The focus then shifted to formal proofs: tactics (*simpl, rewrite, induction, apply*) for proving function properties were demonstrated, to show the transition from verification by example to proof of correctness. Next, through the introduction of type classes and their instances, participants were introduced to the abstraction mechanism for defining algebraic structures (monoids) and polymorphic functions, supported by links to the official documentation and the Software Foundations textbook [41] for independent study. The final part of the tutorial focused on applying Rocq to program verification: using a simple imperative language as an example, the syntax, operational semantics, and state were defined, followed by a demonstration of how to formulate and prove assertions about program behavior, setting the stage for the static analysis hackathon challenge. Methodologically, the tutorial combined structured material, live coding with comments, interactive Q&A sessions, and a focus on the upcoming practical task.

The invited lecture by international expert Alexei Lisitsa (University of Liverpool) deserves special attention; his talk "Finite Countermodel Finding for Infinite State and Parametrized Verification" (based on the prior work [33]) enhanced the scientific value of the event. During the hackathon itself, participants worked on solutions and had the opportunity to consult with mentors via a specially created Telegram-channel, which served as the main means of operational support, announcements, and discussions. The event concluded with an awards ceremony, held both in person and remotely, where the results were summarized and prizes from industrial partners were presented. Therefore, the VeHa design combined self-study, online sessions, intensive practical work, and social interaction. All these different parts of the event created an environment for deep immersion in the field of formal methods.

The organizational experience of previous years, described in the article summarizing VeHa-2024 [31], demonstrated that using a public GitHub repository with a pull-request mechanism for accepting submissions provides high convenience and familiarity for a modern IT audience. This approach allowed participants to work in a familiar ecosystem, easily manage versions of their solutions, and allowed organizers to quickly review code. However, its key drawback was the risk of unauthorized borrowing of ideas and code between teams, as all pull requests became visible after submission, which contradicted the principles of fair competition. In 2025, seeking to improve security and formalize the process, the organizers transitioned to a specialized testing system used for student programming competitions in the style of the ICPC [1]. This testing system, provided by Novosibirsk State University (NSU-TS) [15], involved the isolated submission of final code or proof text via a web interface, followed by a jury review (we did manual reviews of solutions in this competition).

This transition revealed a trade-off between security and usability. On the one hand, the ICPC-like system guaranteed strict solution isolation, ensuring that teams could not see each other's submissions until the end of the competition, thereby enhancing fairness and objectivity. On the other hand, as noted in oral discussions, the interface and workflow of this system proved less flexible and familiar to participants. They lost the ability to maintain a public or private repository with change history within the competition platform. To compensate for the loss of convenience and support collaborative learning, the organizers created a public GitHub repository [2] for the final publication of all reviewed solutions after the competition. The issue of solution submission remains open for future competitions.

4   Hackathon Tasks Design as an Opportunity to Boost FM Education

The VeHa-2025 competition offered participants a set of six tasks covering key areas of modern formal methods. Each problem was designed not only to check technical skills but also to address substantive issues of scientific and practical significance that the organizers encountered in their experience working with formal methods. Teams could select one or more problems from the list.

The first (research) task – proposed by Natalia Garanina, Irina Shoshmina, and Sergey Staroletov – focused on applying model checking methods to the analysis of a catastrophic scenario in the CTCS high-speed train control system ("Wenzhou train collision"). The organizers compiled a rough sequence of the crash in natural language by analyzing open sources (mainly those rooted in the Chinese Wikipedia article about the incident). Participants were required to construct a formal model in the Promela language incorporating equipment fault tolerance, automatic blocking protocols, interactions between dispatch systems and onboard controllers, and time constraints typical of a real-world incident. The key challenge lay in formulating temporal safety properties whose violation leads to a collision. The jury asked for solutions that not only identified a counterexample but also proposed architectural or protocol changes that would guarantee safety in the presence of multiple failures. Among the reading materials, participants were asked to review existing works on the formal analysis and modeling of such incidents [38]. In general, this task continues the disaster modeling approaches used in the first VeHa competition [45].

The second (industrial) task – proposed by Vasil Dyadov and Ilya Shchepetkov – focused on the specification of a distributed version control system using the TLA+ language. Participants were required to formalize the behavior of multiple agents performing parallel file editing, their synchronization via a central server, and conflict resolution algorithms (Push, Pull, Merge). Particular attention was paid to the correct maintenance of a linear change log and guarantees of eventual consistency under diverging histories. The task was structured as a sequence of increasingly complex modules, requiring step-by-step verification of consistency invariants and the absence of data loss using the built-in TLC model checker.

The third (industrial) task – proposed by Alexander Kogtenkov – involved working with the Isabelle/HOL interactive theorem prover and aimed at formally proving properties of functional programs operating on recursive data structures. Participants had to prove a series of lemmas about list folds, the equivalence of different implementations of prefix operators, and properties of functions operating on lists of lists, modeling, for example, manipulaticns of environment tables. The methodological limitation of using only basic tactics (simp, blast, cases, induction) required careful design of proofs and a deep understanding of the principles of structural induction and recursion in HOL.

The fourth (industrial) task – proposed by Timofey Cherganov and Ivan Smirnov – was in the area of static analysis and was implemented in the Rocq/Coq environment. It required the construction and proof of correctness of an abstract interpreter [16] for a simple imperative language. Participants had to define abstract domains (including a flat constant lattice and interval analysis), formalize the lattice structure on states, prove the monotonicity and correctness of abstract operations, and extend the analysis by taking into account information from branching and loop conditions. A key aspect was proving the connection between concrete and abstract semantics through the concretization relation and establishing the preservation of safety properties, such as the absence of division by zero.

The fifth (research) task – proposed by Dmitry Kondratyev – was devoted to the deductive verification of the historical Chakravala algorithm attributed to Indian mathematicians Bhāskara II [10, 19, 42]. This method is used for solving Pell's equation $x^2 - ny^2 = 1$ in integers. The basis of this method is one of the first applications of induction in the history of mathematics. The C implementation of this algorithm contains a nontrivial loop with a large number of interrelated variables, requiring the identification of an invariant reflecting the arithmetic essence of the method. Participants were required to work in the C-lightVer system based on the ACL2 theorem prover [36] and express invariants in Applicative Common Lisp using the provided domain theory. In many cases the ACL2 theorem prover can automatically apply lemmas from domain theory to prove verification conditions generated by the C-lightVer tool. The main lemmas in the domain theory are theorems about recurrent relations between variable values on previous, current and next iterations (these relations have been clearly explained in the Ayyangar's paper [10]). Thus, participants can use these recurrent relations to define loop invariant without losing proof automation. The proof automation is important for students at the first stage of learning Hoare logic where the stage of proving verification conditions is temporarily unnecessary. The challenge lay in defining invariant for loop from the implementation of Chakravala algorithm, which required a deep understanding of the mathematical base of the method.

The sixth and final (research) task – proposed by Vladimir Shelekhov – is taken from a Pascal programming textbook from the 1980s. The task is to select the longest substring of decimal digits that is closest to the left. This seemingly simple string search task was offered to VeHa 2025 participants. It was necessary to write an effective program, build a formal specification of the program (including pre- and post-conditions, loop invariants, and result correctness criteria) and conduct its full deductive verification using a tool of their choice (Why3 [20], Coq or others). An expert assessment of effectiveness is an additional goal of the competition.

5   Results and Analysis

5.1   Analysis of the Competencies Acquired by Competition Participants

Let us analyze the solutions sent to the jury for review (now posted in the repository [2]). The last two are given with reference to the code.

Task 1. The only team that submitted a solution demonstrated average mastery of competencies in verifying complex distributed systems. The participants demonstrated familiarity with abstract domain modeling in the Promela language, correctly identifying the main components of the train control system (TCC, ground equipment, signaling system, ATP, on-board computers, dispatching service, and train processes) and formalizing their interactions via asynchronous data channels. They applied the principles of specifying parallel processes taking into account the state of the system, including modeling block section occupancy, automatic interlock operation, and transitions between control modes. The team also demonstrated the ability to formulate safety requirements in the LTL language, defining a collision-free invariant, and integrating it into the model code for subsequent verification. An important aspect was the modeling of equipment failures and the system's response to them, demonstrating an understanding of the need to consider abnormal scenarios during verification.

Task 2. While solving the task of developing a specification for a distributed version control system in the TLA+ language, the hackathon participants successfully applied abstract modeling to describe the system state and learned to formally define invariants and temporal properties that ensure the correct operation of a distributed algorithm. By iteratively increasing the complexity of the specification, the participants mastered the principles of compositional design, moving from simple file and user models to complex synchronization mechanisms, including conflict handling through Push, Pull, and Merge operations. A significant achievement was their ability to work with the limitations of the TLC model checker, optimizing the state space and adjusting invariants to account for the specifics of merge operations. Their solutions also demonstrate an understanding of distributed systems principles, such as log consistency, linearity of change history, and the guarantee of data preservation during synchronization.

Task 3. Hackathon participants demonstrated basic formal verification competencies in Isabelle/HOL, successfully mastering the definition of primitive recursive functions and proving their properties through structural induction and case analysis. They learned to correctly formulate specifications in a higher-order language, work with standard data structures (lists and lists of lists), and apply key proof methods such as induction on function arguments and case analysis. Participants also demonstrated the ability to construct auxiliary lemmas, relate different function representations (for example, the library 'take' and their own 'pick'), and prove their equivalence. This shows an understanding of the principles of modularity and reuse in formal development. Despite some incomplete fragments, the overall level of their solutions reflects well-developed skills in logical thinking, working with interactive proofs, and applying formal methods to data structure processing problems.

Task 4. While solving problems on implementing a correct static analyzer based on abstract interpretation, the hackathon participants demonstrated a confident grasp of abstract domain theory, including formalizing lattices, defining join, meet, and order operations, and constructing monotone concretization mappings linking abstract and concrete values. Participants successfully applied the apparatus of abstract interpretation in practice, implementing the semantics of expressions and commands, proving its correctness with respect to concrete semantics using inductive methods and lemmas in Coq. Of particular note were their ability to work with complex data structures, such as finite state maps, and correctly define lattice operations on them. Conditional analysis techniques were mastered, including narrowing abstract states using *assume_true* and *assume_false*, which improved the accuracy of analysis for conditional statements and loops. Participants also gained experience integrating various abstract domains (flat values and intervals) and adapting them to the specifics of the problem. During their work, the teams encountered typical verification problems, such as proving fixed-point convergence and the correctness of loop analysis.

Task 5. While developing solutions for the deductive verification of the Chakravala algorithm, the participants successfully applied their deep understanding of the algorithm's mathematical foundations, formalizing complex recurrence relations and integer properties of variables as loop invariants in the ACL2 language. Their ability to work with the C-lightVer system and integrate domain-specific theorems from the applied theory demonstrates their mastery of automated verification tools. Furthermore, participants demonstrated skills in analyzing and structuring invariants, taking into account various loop execution states, such as initialization, iteration, and termination, and used debugging practices to verify the correctness of the implementation. These competencies reflect the ability to combine theoretical knowledge with practical application in the context of verifying knowledge-intensive software. The loop invariant proposed by the task author can be found in "chakravala.c" file [29] as a content of the comment before loop. Let us consider an important part of the loop body:

```
if (f == 1){
    j = s; g = h; h = a; u = v; v = b; w = m + j;
    c = k; z = w / c; a = h*z; b = v*z; a = a - g; b = b - u;
    k = z*w; d = 2*j; d = d*z; k = k - d; t = j*j; t = t - n;
```

```
    t = t / c; k = k + t;}
```

This part of loop body depends on the variables a, b, k, m, g, h, u, v, c, j, f. If $a^2 - nb^2 = k$ is Pell's equation with parameter k then g stores a value from the previous iteration, h stores a value from the current iteration, a stores a value from the next iteration, u stores b value from the previous iteration, v stores b value from the current iteration, b stores b value from the next iteration, c stores k value from the current iteration, k stores k value from the next iteration, m is the next value connecting k with c, j stores the m value from the current iteration, and f indicates whether a previous m value exists. Let us consider defining, in the invariant, the recurrence relations that correspond to this part of loop body. The participants' solutions are close to the author's solution. Thus, let us consider this part of invariant in the author's solution:

```
(implies
    (and (= f 1) (not (= c 0)))
    (and
        (= g (/ (- (* h j) (* v n)) c)) (= u (/ (- (* v j) h) c))
        (= h (/ (- (* a m) (* b n)) k)) (= v (/ (- (* b m) a) k))
        (= a (/ (+ (* h m) (* v n)) c)) (= b (/ (+ h (* m v)) c))
        (= k (/ (- (* m m) n) c)))))
```

In solutions, students have demonstrated the ability to define these recurrence relations derived from the Chakravala method.

Task 6. Analyzing the solutions presented by hackathon participants, it can be noted that, overall, participants acquired practical skills in formalizing problem statements, constructing specifications using preconditions and postconditions, and writing loop invariants. They mastered working with modern verification tools such as Why3, Coq (Rocq), and Lean [17], including both automated proofs using SMT solvers and interactive proof generation. Furthermore, participants developed the ability to critically evaluate solutions, identifying errors in specifications (like unnecessary constraints or incorrect axioms) and understanding the importance of precise formal formulations. The ability to generalize problems by moving from a concrete formulation (working with numbers) to abstract models using uninterpreted predicates was also demonstrated.

When building the program, the participants used one of two ways to represent the source string: as a classical list in the style of the Haskell language [26] or as an array. It turned out that representing a string as a list is much more difficult for specification and deductive verification. The specification of a program with a list representation can be defined by the following predicate in the WhyML language [20]:

```
predicate maxLeftSubstring(s d: string)(n: int) =
 maxSubstring s d n /\
 exists u v: string. s = u ++ d ++ v /\
  forall d1: string, m: int. maxSubstring u d1 m -> n = 0 \/ m < n
```

Here, the parameter **d** is the leftmost decimal substring of length **n** in the original string **s**. The predicate **maxSubstring s d n** is a statement that **d** is the maximum decimal substring of length **n**. In the predicate **maxLeftSubstring**, the leftmost maximality is guaranteed by the fact that to the left of the substring **d**, in the substring **u**, all its decimal substrings are less than **n** in length. The disjunction **n = 0** is necessary here. This specification is not trivial. Verification also turned out to be difficult, since the presence of existential quantifiers makes automatic verification of correctness formulas problematic. For the given specification via the **maxLeftSubstring** predicate, the proof required the introduction of three non-trivial lemmas.

The specification and verification using a list representation of the source string, implemented by the participants in the Rocq tool, also appeared difficult. Several participants who chose a list representation failed to complete the task. On the contrary, the participants who chose the Why3 tool with an array representation successfully completed the task. The specification was written without the use of existential quantifiers, as a result of which the correctness of formulas was checked automatically using SMT solvers. All the programs presented by the participants had significant efficiency defects. The task of building and verifying the most effective program turned out to be very difficult [43].

5.2 Analysis of Team Registrations and Competition Results

Quantitative participation figures for VeHa-2025 clearly demonstrate both overall engagement and participants' tool preferences. Of the 35 teams registered via Google Forms (47 participants from 8 cities), 19 teams, comprising 30 participants from 4 cities, progressed to the active problem-solving phase. The statistics on submitted solutions by tool revealed a clear preference for deductive verification (Fig. 2). The Isabelle/HOL tool was the clear leader in the number of solutions (8 solutions), followed by TLA+ (7 solutions) and Rocq (Coq) (3 solutions). Two solutions each were submitted using Why3 and C-lightVer, while only one solution each was submitted using SPIN/Promela model checking tools and the Lean/Loom/Velvet combination (the latter problem was clearly not limited to a tool). This distribution, where deductive methods (Isabelle/HOL, Coq, Why3, C-lightVer) accounted for 15 solutions versus 8 solutions for model checking methods (TLA+, SPIN), reflects both the higher entry barrier to full-fledged modeling of complex systems and the greater prevalence or availability of training materials on deductive tools in the academic community.

The most impressive results were achieved by teams from St. Petersburg, which were not only the largest but also the most productive. For example, the "VeryVerifiers" team (SPbPU) demonstrated the breadth of their expertise, receiving awards in three different categories: a certificate for TLA+, a third-place diploma for Isabelle/HOL, and a second-place diploma for verification of a search algorithm. The "SPbPU-40202" team became champion on the Chakravala algorithm track and received a certificate for Isabelle/HOL. The "Pleshcheevo Ozero Surf Club" team (Bauman Moscow State Technical University, Moscow) deserves special attention, winning a second-place TLA+ diploma

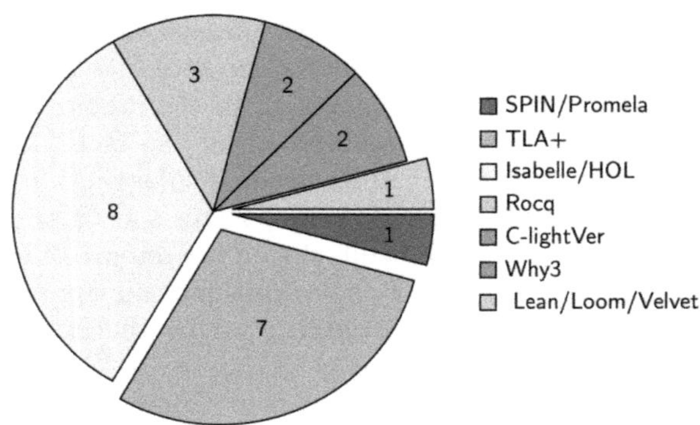

Fig. 2. Number of submitted solutions by verification tools.

and a certificate in Coq, having won two nominations from different industrial partners. Among individual achievements, Ivan Smirnov (MIPT/MIAN/Astra Linux) stands out, receiving awards for both search verification and his work with TLA+. The exceptional complexity of the Chinese train collision problem (SPIN/Promela) meant that only one team, called T1, partially modeled it and received a special diploma.

The geographic distribution of the winners confirmed the emergence of strong national centers of excellence in formal methods. St. Petersburg was the clear leader in the total number of awards, with teams (primarily from SPbPU and ITMO) taking home 14 awards, having solved the largest number of problems overall, including several teams that solved three problems each. Novosibirsk, represented by teams from Novosibirsk State University and SB RAS institutes, took second place with seven awards, demonstrating particularly strong results in the Isabelle/HOL and TLA+ tracks. Moscow, with active participation from Bauman Moscow State Technical University, Moscow Institute of Physics and Technology, and the Institute of System Programming of the Russian Academy of Sciences, received five awards. Also noteworthy is the international participant from the University of Naples (Paphos, Cyprus), who took first place in the category "Verification of an Efficient Search Algorithm" (from the team of Loom/Velvet [22] developers). Thus, the VeHa-2025 results not only identified the current leaders but also mapped the geographic landscape of the community, showing us possible centers for scientific collaboration.

5.3   Feedback Analysis

Based on the feedback received from 12 anonymous participants of our formal verification hackathon (collected via Google Forms), we conclude that the event was highly rated overall: 67% of respondents (8 out of 12) gave the maximum rating (5.0) to the overall level of organization. All respondents characterized the level of the proposed tasks as "okay" and their wording as "fairly adequate". Furthermore, 42% of participants (5 out of 12) expressed a desire for a longer event.

Regarding the format, 67% would prefer a hybrid (in-person/online) option, while the remainder preferred a fully online format. Key areas for improvement identified in the feedback included a desire to segment tasks into time blocks and improve the solution submission system – specifically, a return to the closed repository model to improve teamwork and control over submitted materials. Furthermore, the importance of familiarizing participants with the solution evaluation criteria in advance was noted. Ratings for specific tasks varied, with some tasks (for example, task 4 on the Rocq) receiving exceptionally high ratings (three 5.0 ratings from three participants).

5.4 Using Data of the VeHa Series for Analytics

All parties involved in the contest can potentially find answers to their questions from the data collected during the contests. Universities (or even the ministry of education) can see whether the curriculum and the level of the material in different cities is comparable. Companies can pick universities that provide skills more suitable for the companies' tasks. Students can find the most efficient organizational unit of participation.

Starting with VeHa-2025, we are developing a framework that would provide such analytics. This involves multiple steps, including legal matters, careful registration of information reported by participants and the jury, unification of the results, etc. In the end it might be possible to figure out if there is any correlation between the following characteristics: participants unit (team or solo), location, participants sex, affiliation kind (a university or a company), results, nomination category (such as model checking vs. deductive verification), difficulty, etc.

We analyzed several hypotheses using the data from 2023 and 2025 and have the following preliminary non-trivial observations:

- Results of participants do not depend on their affiliation, including both location and organizational form (a university or a company).
- Hard tasks are correctly identified by the authors, i.e. are solved by a smaller number of participants with lower results. With easy and medium-difficulty tasks the difference is not always obvious.
- On average, teams with participants of both sexes perform better than any other team composition.
- Participants from some locations prefer teamwork while participants from other locations prefer solo.
- Participants from one specific location perform better in deductive verification, staying at the average level in model checking.

While this information is interesting, drawing conclusions about why a given correlation exists is currently beyond our scope, as it would require setting up experiments to confirm or refute hypotheses about potential causation.

6 Conclusions and Future Work

As a result, we can conclude that the third verification hackathon, VeHa-2025, confirmed the effectiveness of the hackathon format as a pedagogical

Fig. 3. Award ceremonies for winners in different cities (Moscow, Novosibirsk, St. Petersburg).

tool in the field of formal methods. The event successfully implemented its key educational principles: hands-on learning through problems, direct interaction between academia and companies seeking specialists in formal methods, and a hybrid study of the two main verification paradigms – model checking and deductive reasoning.

The quantitative and qualitative results presented in the article confirm that our event is on the right track. Despite the relatively high complexity of the problems, the established community of participants from leading universities and research centers (St. Petersburg, Novosibirsk, and Moscow) demonstrated their ability to solve all of these problems, which are unusual for the average student developer and involve formal modeling of critical systems, specification of distributed algorithms, and proof of program correctness. In Fig. 3 we show the awards ceremonies held in these cities after the event. The successful completion of problems developed by our industry partners and the positive feedback from participants highlight the relevance of this competitive format for training exceptional specialists in this specific field.

While the diversity of verification tools used by participants in our competition may provide a broad introduction to formal methods, it also raises the question of whether such a large number of languages and tools might be overwhelming for students, especially those with limited experience. However, the current design of our competition initially assumes that participants will choose the track/task they are interested in pursuing before the competition. After the full problem statements are announced and the tutorials are completed, they may change their minds or even submit solutions for multiple tracks. Typically, each task was associated with a specific verification tool, but for the final task, the tool was not explicitly specified, which attracted diverse communities to participate and even led to submissions in languages unexpected by the organizers. Therefore, skill development was not limited to the participants.

Our experience also revealed several important areas for improvement. First, there is a need to further balance the task difficulty. We see that tasks with multiple difficulty levels are generally better solved. Participants may be discouraged by tasks with fuzzy formulations during their initial selection. Second, we need to consider optimizing the solution submission infrastructure and improving integration with the competition website, including better accessibility of materials and links to third-party training courses. Third, we would like to attract more third-party participants, including from companies, not just research centers affiliated with the competition organizers. We are also considering a more in-depth analysis of all submitted solutions, which allows us to create a structured database in the form of benchmarks for various verification systems, including those being developed at local universities. We also recognize that since the competition is organized in our free time slots, full integration of all training tools into a single system is lacking.

Another related issue is the problem of attracting students to research in the foundations of formal methods. The idea of launching VeHa-2023 came as a booster for the annual workshop on Program Semantics, Specification, and Verification (PSSV). It was assumed that students participating in VeHa would also attend (at least) invited talks at PSSV and become "engaged" in advanced theories. However, in reality, none of the VeHa-2023 or VeHa-2024 participants attended either PSSV-2023 or PSSV-2024. In contrast, during VeHa-2025, the invited lecture by Alexei Lisitsa held online deserved special attention and high attendance. Perhaps this format (a single invited lecture) is much better than a workshop running in parallel.

Finally, the VeHa competition series has established itself as a sustainable platform for the popularization, education, and practical application of formal methods through example. Continuing this initiative, developing international collaboration, and strengthening ties with industry will contribute to the further growth of the community and increase the reliability of the software being developed.

Acknowledgements. The authors would like to thank our industrial partners, Kaspersky Lab and Astra Group, for their support in proposing practical-oriented tasks and providing the prizes. We are also grateful to the Ivannikov Institute for System Programming of the Russian Academy of Sciences (ISP RAS), especially A.K. Petrenko, for organizing the award ceremony for VeHa winners and assistance in publishing the analytics papers about previous competitions.

References

1. ICPC Foundation. https://icpc.foundation
2. VeHa-2025 Solutions. https://github.com/VeHaContest/VeHa2025
3. Yandex Telemost. https://360.yandex.ru/telemost/?ncrnd=881
4. Codeforces (2009–2026). https://codeforces.com
5. Workshop of Program Semantics, Specification and Verification: Theory and Applications (2023). https://persons.iis.nsk.su/en/PSSVfrom2022towards2023
6. Workshop of Program Semantics, Specification and Verification: Theory and Applications (2024). https://persons.iis.nsk.su/ru/PSSV-2024
7. VeHa-2025 (2025). (in Russian). https://sites.google.com/view/veha2025/

8. Huisman, M., Monti, R., Ulbrich, M., Weigl, A.: The VerifyThis Collaborative Long Term Challenge. In: Ahrendt, W., Beckert, B., Bubel, R., Hähnle, R., Ulbrich, M. (eds.) Deductive Software Verification: Future Perspectives. LNCS, vol. 12345, pp. 246–260. Springer, Cham (2020). https://doi.org/10.1007/978-3-030-64354-6_10

9. Ahrendt, W., Herber, P., Huisman, M., Ulbrich, M.: SpecifyThis-bridging gaps between program specification paradigms. In: International Symposium on Leveraging Applications of Formal Methods, pp. 3–6. Springer (2022)

10. Ayyangar, A.K.: New light on Bhaskara's Chakravala or cyclic method of solving indeterminate equations of the second degree in two variables. J. Indian Math. Soc. **18**, 225–248 (1929–30)

11. Baudin, P., et al.: The dogged pursuit of bug-free C programs: the Frama-C software analysis platform. Commun. ACM **64**(8), 56–68 (2021). https://doi.org/10.1145/3470569

12. Bertot, Y.: A Short Presentation of Coq. In: Mohamed, O.A., Muñoz, C., Tahar, S. (eds.) TPHOLs 2008. LNCS, vol. 5170, pp. 12–16. Springer, Heidelberg (2008). https://doi.org/10.1007/978-3-540-71067-7_3

13. Beyer, D., Huisman, M., Kordon, F., Steffen, B.: TOOLympics II: competitions on formal methods. Int. J. Softw. Tools Technol. Transfer **23**(6), 879–881 (2021). https://doi.org/10.1007/s10009-021-00631-1

14. Chas, K.: The world's smartest programmers compete: ACM ICPC. Commun. ACM (2013). https://cacm.acm.org/blogs/blog-cacm/165692-the-worlds-smartest-programmers-compete-acm-icpc/

15. Churina, T., Nesterenko, T.: 20 years of experience in organizing and holding of programming competitions at the Novosibirsk State university. In: 2020 Fifth International Conference "History of Computing in the Russia, former Soviet Union and Council for Mutual Economic Assistance countries" (SORUCOM), pp. 139–144. IEEE (2020)

16. Cousot, P.: Abstract interpretation. ACM Comput. Surv. (CSUR) **28**(2), 324–328 (1996)

17. de Moura, L., Kong, S., Avigad, J., van Doorn, F., von Raumer, J.: The Lean Theorem Prover (System Description). In: Felty, A.P., Middeldorp, A. (eds.) CADE 2015. LNCS (LNAI), vol. 9195, pp. 378–388. Springer, Cham (2015). https://doi.org/10.1007/978-3-319-21401-6_26

18. Denis, X., Siegel, S.F.: VerifyThis 2023: an international program verification competition. In: TOOLympics Challenge 2023, pp. 147–159. Springer (2025)

19. Dutta, A.K.: Kuṭṭaka, Bhāvanā and Cakravāla. In: Seshadri, C.S. (ed.) Studies in the History of Indian Mathematics. CHM, pp. 145–199. Hindustan Book Agency, Gurgaon (2010). https://doi.org/10.1007/978-93-86279-49-1_7

20. Filliâtre, J.-C., Paskevich, A.: Why3 — Where Programs Meet Provers. In: Felleisen, M., Gardner, P. (eds.) ESOP 2013. LNCS, vol. 7792, pp. 125–128. Springer, Heidelberg (2013). https://doi.org/10.1007/978-3-642-37036-6_8

21. Geske, M., Jasper, M., Steffen, B., Howar, F., Schordan, M., van de Pol, J.: RERS 2016: Parallel and Sequential Benchmarks with Focus on LTL Verification. In: Margaria, T., Steffen, B. (eds.) ISoLA 2016. LNCS, vol. 9953, pp. 787–803. Springer, Cham (2016). https://doi.org/10.1007/978-3-319-47169-3_59

22. Gladshtein, V., Pîrlea, G., Zhao, Q., Kurin, V., Sergey, I.: Foundational multimodal program verifiers. Proc. ACM Program. Lang **10**, 2233–2264 (2026). https://doi.org/10.1145/3776719 POPL

23. Gorodnyaya, L.V., Irtegov, D.V., Nepeivoda, N.N., Pottosin, I.V., Churina, T.G.: All-siberian open olympiad in programming (Novosibirsk State Univer-

sity). Program. Comput. Softw. **27**(3), 165–167 (2001). https://doi.org/10.1023/A:1010942517538

24. Holzmann, G.J.: The model checker SPIN. IEEE Trans. Software Eng. **23**(5), 279–295 (1997)

25. Howar, F., Jasper, M., Mues, M., Schmidt, D., Steffen, B.: The RERS challenge: towards controllable and scalable benchmark synthesis. Int. J. Softw. Tools Technol. Transfer **23**(6), 917–930 (2021). https://doi.org/10.1007/s10009-021-00617-z

26. Hutton, G.: Programming in Haskell, Cambridge University Press (2016)

27. Jasper, M., Mues, M., Murtovi, A., Schlüter, M., Howar, F., Steffen, B., Schordan, M., Hendriks, D., Schiffelers, R., Kuppens, H., Vaandrager, F.W.: RERS 2019: Combining Synthesis with Real-World Models. In: Beyer, D., Huisman, M., Kordon, F., Steffen, B. (eds.) TACAS 2019. LNCS, vol. 11429, pp. 101–115. Springer, Cham (2019). https://doi.org/10.1007/978-3-030-17502-3_7

28. Khazeev, M., Aslam, H., de Carvalho, D., Mazzara, M., Bruel, J.-M., Brown, J.A.: Reflections on Teaching Formal Methods for Software Development in Higher Education. In: Bruel, J.-M., Capozucca, A., Mazzara, M., Meyer, B., Naumchev, A., Sadovykh, A. (eds.) FISEE 2019. LNCS, vol. 12271, pp. 28–41. Springer, Cham (2020). https://doi.org/10.1007/978-3-030-57663-9_3

29. Kondratyev, D.: Deductive verification of a program for solving Pell's quadratic Diophantine equation implementing the Chakravala method developed in India, which is one of the first cyclic algorithms in the history of mathematics (2026). https://github.com/VeHaContest/VeHa2025/tree/main/task5_c-lightver/Dmitry_Kondratyev

30. Kondratyev, D.A., Nepomniaschy, V.A.: Automation of C program deductive verification without using loop invariants. Program. Comput. Softw. **48**(5), 331–346 (2022). https://doi.org/10.1134/S036176882205005X

31. Kondratyev, D.A., et al.: VeHa-2024 formal verification contest: two years of experience and prospects. Trudy ISP RAN/Proc. ISP RAS **37**(1), 159–184 (2025). (in Russian)

32. Lamport, L., Matthews, J., Tuttle, M., Yu, Y.: Specifying and verifying systems with TLA. In: Proceedings of the 10th Workshop on ACM SIGOPS European Workshop, pp. 45–48. (2002)

33. Lisitsa, A.: Finite model finding for parameterized verification, (2010). arXiv preprint

34. Liu, A.: Deepseek-v3 technical report, (2024). arXiv preprint

35. Moniz, H.: The Istanbul BFT consensus algorithm. CoRR abs/2002.03613 (2020). https://arxiv.org/abs/2002.03613

36. Moore, J.S.: Milestones from the Pure Lisp theorem prover to ACL2. Formal Aspects Comput. (6), 1–34 (2019). https://doi.org/10.1007/s00165-019-00490-3

37. Natarajan, V., Holzmann, G.J.: Outline for an operational semantics of Promela. DIMACS Ser. Discrete Math. Theor. Comput. Sci **32**, 133–152 (1997)

38. Ouyang, M., Hong, L., Yu, M.H., Fei, Q.: STAMP-based analysis on the railway accident and accident spreading: taking the China-Jiaoji railway accident for example. Saf. Sci. **48**(5), 544–555 (2010)

39. Paulin-Mohring, C.: Introduction to the Coq Proof-Assistant for Practical Software Verification. In: Meyer, B., Nordio, M. (eds.) LASER 2011. LNCS, vol. 7682, pp. 45–95. Springer, Heidelberg (2012). https://doi.org/10.1007/978-3-642-35746-6_3

40. Paulson, L.C., Nipkow, T., Wenzel, M.: From LCF to Isabelle/HOL. Formal Aspects Comput. (6), 1–24 (2019). https://doi.org/10.1007/s00165-019-00492-1

41. Pierce, B.C., et al.: Software foundations (2018). https://idris-hackers.github.io/software-foundations/pdf/sf-idris-2018.pdf
42. Selenius, C.O.: Rationale of the Chakravāla process of Jayadeva and Bhāskara ii. Hist. Math. **2**(2), 167–184 (1975). https://doi.org/10.1016/0315-0860(75)90143-3
43. Shelekhov, V.: Task 6. Solving via Cycles. Materials of the VeHa 2025 formal verification competition. Preprint 213, A.P. Ershov Institute of Informatics Systems, SB RAS, Novosibirsk, Russia, 13p. (2025). (in Russian). https://www.iis.nsk.su/files/preprint/leftsub.pdf
44. Staroletov, S.: Teaching the discipline "software testing and verification" to future programmers. Syst. Inform. (21), 1–28 (2022). https://doi.org/10.31144/SI.2307-6410.2022.N21.P1-28
45. Staroletov, S.M., Kondratyev, D.A., Garanina, N.O., Shoshmina, I.V.: VeHa-2023 formal verification contest: the experience. Trudy ISP RAN/Proc. ISP RAS 36(2), 141–168 (2024). (in Russian). https://doi.org/10.15514/ISPRAS-2024-36(2)-11
46. Yu, Y., Manolios, P., Lamport, L.: Model checking TLA+ specifications. In: Advanced Research Working Conference on Correct Hardware Design and Verification Methods, pp. 54–66. Springer (1999)
47. Zeng, Y., Lin, F., Tian, L., Gan, J., Chen, Z.: IBFT: an impartial Byzantine fault tolerance consensus protocol for blockchain. Blockchain Res. Appl , 100430 (2025). https://doi.org/10.1016/j.bcra.2025.100430

Author Index

GPSR Compliance
The European Union's (EU) General Product Safety Regulation (GPSR) is a set
of rules that requires consumer products to be safe and our obligations to
ensure this.

If you have any concerns about our products, you can contact us on

ProductSafety@springernature.com

In case Publisher is established outside the EU, the EU authorized
representative is:

Springer Nature Customer Service Center GmbH
Europaplatz 3
69115 Heidelberg, Germany